The New Global Puzzle
What World for the EU in 2025?

Directed by Nicole Gnesotto and Giovanni Grevi

Institute for Security Studies
European Union
Paris

Institute for Security Studies
European Union
43 avenue du Président Wilson
75775 Paris cedex 16
tel.: +33 (0)1 56 89 19 30
fax: +33 (0)1 56 89 19 31
e-mail: info@iss.europa.eu
www.iss.europa.eu

Director: Nicole Gnesotto

© EU Institute for Security Studies 2006. All rights reserved. No part of this publication may be reproduced, stored in a retrieval system or transmitted in any form or by any means, electronic, mechanical, photocopying, recording or otherwise without the prior permission of the EU Institute for Security Studies.
ISBN 92-9198-096-X

Published by the EU Institute for Security Studies and printed in Condé-sur-Noireau (France) by Corlet Imprimeur. Graphic design by Claire Mabille (Paris). Cover design by Eurofuture.

Contents

The Team 5

Introduction 7

Part I: The Trends

1. Demography 15
2. Economy 31
3. Energy 53
4. Environment 75
5. Science and Technology 91

Part II: The Regions

6. Eurasia and Russia 103
7. The Middle East and North Africa 117
8. Sub-Saharan Africa 131
9. The United States 141

Contents...

10	China	155
11	India	165
12	Latin America	177

The EU in Context

The world in 2025: a snapshot	189
Three defining questions for the future	195
The European Union in context	203

Annexes
- *Bibliography* — 215
- *Abbreviations* — 249

The Team

This Report is the result of a collaborative effort involving, under the leadership of Director **Nicole Gnesotto**, the research team of the Institute and a group of Visiting Fellows and Research Assistants whose added value to the work of the Institute cannot be stressed enough.

Burkard Schmitt, Assistant Director at the Institute, conducted the first stage of the Long-Term Vision project, between the end of 2005 and March 2006, with the involvement of Giovanni Grevi and Gustav Lindstrom from among the Research Fellows. His input and leadership were decisive in defining the scope of the exercise, drafting the first version of the Report to the EDA, as well as managing the team. He subsequently left the Institute for a position in the European Commission, but his name will remain associated with this publication.

Giovanni Grevi, Research Fellow at the Institute, took charge of the second stage of the exercise: without his full-time commitment and dedication, this project would never have been completed. He was responsible for thoroughly reviewing and expanding the first version of the Report, while managing a team of Research Assistants, and drafting a wide range of new sections. He shared the final editing of the book with the Director.

Most of the Research Fellows of the Institute were also directly involved in this project. Their distinctive contribution provided much added value by bringing in specific regional expertise. In particular, they highlighted some of the defining questions affecting the future internal and external developments of key countries and regions. In this context, special mention should be made of their work, including:

Dov Lynch on Russia and Eurasia;
Walter Posch on the Middle East and the Mediterranean;
Pierre-Antoine Braud on Sub-Saharan Africa;
Marcin Zaborowski on the United States;
Martin Ortega on Latin America;
and **Gustav Lindstrom** on Science and Technology.

A team of Research Assistants worked throughout the spring and summer of 2006 collecting and reviewing relevant sources, checking, completing, and analysing all the facts and figures that form the basis of the Report and providing their own insight and advice on a number of issues.

Stéphane Delory has been involved as Research Assistant since the very beginning of the project and his key contribution deserves a special mention. **Vincent Vicard**, who joined the team in spring, was responsible for the economic dimension of the study. **Christian Burckhardt**, **Johan Edqvist**, **Sophie de Laboulaye**, **Bastien Nivet**, **Patrizia Pompili**, **Andras Roth**, **Nathalie Stanus** and **Daniel Steinvorth** also participated in the project. Last but not least, the Institute's English-language editor, **Gearóid Cronin**, did an excellent job in polishing up the final draft of the Report.

Introduction

Nicole Gnesotto

The background to the project

At the end of 2005, the European Defence Agency (EDA) invited the European Union Institute for Security Studies to participate in the development of an initial Long-Term Vision (LTV) study for European capability and capacity needs in the field of ESDP. The task of the EUISS was to prepare, on the basis of existing relevant studies, a report on the global context in which the European Security and Defence Policy will operate in 2025. Our mission was not to foretell the future but to assess the most likely developments in several key areas, on the basis of existing open sources. Other EU bodies were involved in this project, addressing the military, industrial and technological dimensions of the Long-Term Vision, under the leadership of the EDA.

The Institute completed its work and delivered a 'Global Context Study' to the EDA in March 2006. Since then, the EDA has been working on the elaboration of a comprehensive Long-Term Vision for future ESDP needs. The Agency will present the final output of the project to Member States under its sole responsibility.

For our part, we felt that the EUISS contribution to the EDA-led project provided a good basis on which to deepen and expand the Institute's reflection on the actors and factors shaping the global context of European integration. The resulting report, *The New Global Puzzle – What World for the EU in 2025?*, in no way represents the official position of the European Union on any specific subject. Nor does it engage any Member State to endorse its findings and conclusions. As is the case for all the other publications of the Institute, an independent think-tank at the service of CFSP, the contents of this publication are the sole responsibility of the Institute.

The structure of the Report

The first part of this Report analyses major long-term trends in five key areas, namely demography, the economy, energy, the environment and science and technology. Since most of these trends differ considerably between the various parts of the world, the general analysis is complemented, where appropriate, by a more detailed examination by region. On the basis of this work, in the second part of the Report the Institute's team presents seven regional outlooks. These combine the main facts and figures highlighted in the first part of the Report with an evaluation of the key political factors influencing the future evolution of major regions and countries. Separate sections belonging to the first and the second parts of the Report have been drafted with a view to being stand-alone. It should be noted that this approach inevitably entails a degree of repetition between the thematic and regional sections. The last part of the Report includes a broader perspective on the state of the international system by 2025, highlights some key political questions and focuses on the EU itself.

The purpose of the Report

This Report does not aim at predicting what the world (and the EU) will look like twenty years from now – an impossible task. Rather, the Institute has undertaken a wide-ranging exercise whose aim is to identify the long-term trends, factors and actors shaping the global environment of European integration — *The New Global Puzzle*. The purpose of this project consists in providing the European public and decision-makers with a better understanding of the variables influencing the evolution of the international system in which the EU will have to act.

Too often European integration is regarded as an insular process, advancing or stalling depending on endogenous, intra-European variables only. Absorbed in endless debates on the reform of their political, economic and social structures, there is a risk that the EU and its Member States may lose sight of the momentous developments taking place beyond their borders, yet affecting their own future. This Report advances a strong plea for the debate on the integration of Europe to be set both 'in context' and 'in perspective'.

As such, this Report does not focus on the EU itself, nor on the evolution and destination of the process of continental integration. The future of the Union is uncertain: how many Member States will have joined the EU by 2025? What level of cooperation and integration will be achieved and in which areas? The answers to these questions will be crucial for Europe's capacity to cope with the political, economic, demographic, energy and security challenges that lie ahead. However, focusing on global developments across a broad spectrum of different variables helps change perspectives and provides new insights. The Report paves the way for further reflection and provides a contribution to strategic policy-making at the European level.

The method

This Report is based on the review of hundreds of documents, including specialised reports, in-depth studies, essays and articles from international organisations, the private sector, academia, think tanks and NGOs. The Institute's team has also benefited from the input of several experts who were consulted concerning specific thematic and geographical issues. Having cross-checked relevant sources, identified and addressed possible inconsistencies and distilled the key questions emerging in each issue area, the Report provides an extrapolation of ongoing trends, with a horizon of 2025.

The assumption of continuity underlying the Report is deliberate. Needless to say, the possibility of major disruptions, changing the course of history, cannot be dismissed. The 'surprise factor', be it the spread of global pandemics, major terrorist attacks, or natural catastrophes, needs to be integrated into strategic policy-making. Even globalisation, the overarching trend affecting almost all others, could conceivably be reversed. Having said that, for the purposes of this exercise, we felt that the assumption of continuity was the only viable one.

In particular, we have decided not to produce alternative scenarios, which are often included in prospective studies to draw attention to different possible futures. This does not mean, however, that we have downplayed uncertainties. On a number of issues, we acknowledge the potential for deviation from established forecasts. In conducting our study, we kept some cautionary guidelines firmly in our minds, so as to avoid sliding into determinism.

All trends are highly interrelated, which means that unexpected developments occurring in one of them will inevitably affect the others. The evolution of most trends depends, among other factors, on political decisions, which can be reversed (e.g. the declining use of nuclear energy in Europe). Last but not least, potential solutions to some of the major problems of our age might stem from ground-breaking innovations whose full impact as yet cannot be foreseen (e.g. progress in biotechnology applied to health, or new applications for renewable energy sources).

On top of that come the specific difficulties affecting each trend area. In some cases, projections are based on assumptions and data which are the subject of debate and/or are very difficult to collect (e.g. regarding oil and gas reserves or migration flows). Other developments are based on models, which are scientifically controversial (e.g. climate change). As a consequence, figures for the same trends often differ widely between various sources (e.g. the expected increase in temperatures across the world).

Conscious of the many twists that could affect the linear evolution of current trends, we nevertheless believe that this Report provides a solid and comprehensive (although by no means exhaustive) picture of the driving factors that will shape the world to come, of the position of major countries and regions therein, and of the questions that all global actors will need to address.

The future

There are two main reasons why we decided to complete, deepen, and expand the scope and the structure of our initial work, and finally to circulate it as a book published by the Institute. First of all, putting together an overall picture of the world in which the EU will operate over the next two decades filled a vacuum and presented a stimulating challenge. This has been one of the most interesting studies that we have undertaken since the creation of the Institute. Reviewing hundreds of documents on such different issues as climate change, foreign direct investments, oil reserves, the spread of contagious diseases or regional conflicts certainly posed a difficult challenge. But it was also an instructive exercise, in so far as this broader vision obliged us to put in perspective the immediate priorities and challenges of EU external policy. The result of this comprehensive study is certainly not beyond questioning. Most of our findings can be discussed, challenged, and interpreted differently. This is why we thought it useful to distribute them to a wide public audience. At a time when the lines of the debate on the future of Europe are blurred, there can never be too much reflection on the shape of the world to come, and the EU's role in it.

Secondly, we hope that we have set a precedent. In the future, we are convinced that it would be extremely useful for the EU to establish a joint mechanism to monitor developments systematically in key areas and regularly develop a comprehensive vision of long-term global dynamics, with a twenty- to thirty-year horizon. Every five years, at the beginning of each new

European 'legislature' (the new terms of office of the European Parliament, the Commission and the High Representative for CFSP/future Union Minister for Foreign Affairs), such joint analysis of the long-term global environment would constitute, in our opinion, an important contribution enabling all EU institutions to behave more strategically, in synergy. This is indispensable if the Union is to face up to its growing responsibilities as a global actor in a challenging global environment.

Part I
The Trends

Demography

The growth of the world's population will slow down as the twenty-first century progresses, due to a general shift in demographic patterns from high birth rates, high mortality and short life expectancy to low birth rates, low mortality and long life expectancy.[1] However, this transition will be incremental, and will take place in several phases, at different times and at different paces in the different parts of the world. Whereas the developed countries are already at an advanced stage of demographic transition, most developing countries are still in the early stages, showing considerably higher birth and population growth rates. Twenty years from now, nine out of ten people will live in the developing world, as opposed to eight out of ten today.

General Trends

- The world population is expected to increase from 6.4 billion in 2005 to 7.9 billion in 2025 (+23.4%). Population growth will be particularly strong in developing countries in Sub-Saharan Africa (+43 to 48.4%), MENA (+38%), Latin America (+24%) and Asia (+21%).

- The total population of the developed world will remain stable, but demographic trends will differ greatly between the US (+17.4%), the EU (+2%), Japan (-2.6%) and Russia (-10.8%).[2] In 2025, the EU-25 will only account for roughly 6% of the world's population.

- Population ageing will be the main demographic feature in developed countries, but also in China.[3] In many EU Member States, but even more so in Japan, this will have major implications for the composition of the workforce and dramatically increase old-age dependency ratios. In many developing countries, in contrast, the population will remain relatively young and the working-age population will expand, putting huge strains on labour markets.

> - Although the number of migrants on the move has been decreasing slightly, the share of international migrants directed towards developed countries has been rising.[4] Migration rates towards Europe are expected to remain stable, while much will depend on the economic performance of neighbouring areas and on European migration policies.
> - Health care will probably continue to improve, but also face new challenges both in the developed world (due to pollution and population ageing) and the developing world (due to pollution and urbanisation). In certain regions, diseases are expected to have a massive impact on the age structure of the population. In general, however, they are unlikely to alter population growth fundamentally. Epidemics will certainly spread more easily due to migration and increased mobility.

By 2025, about 1.2 billion people will be older than 60. In absolute terms, the majority of older people will live in developing countries, but their share of the total population will be much higher in developed countries (30% as compared to 13% in the developing countries).[5] In the long-term, population ageing will lead to a labour force deficit in many developed countries. However, the full effect of this trend will be seen only from 2030 on.

Up until 2025, technological adjustment, the expansion of workforce participation rates and immigration will probably be sufficient to maintain sizeable workforces. The effective old-age dependency ratio (number of non-active persons older than 65 as a percentage of employed persons between 15 and 64), however, will increase dramatically over the next twenty years, in particular in Europe and Japan. Most developing countries, in contrast, will have difficulty generating enough jobs to absorb their labour force stocks.

Because of the many variables affecting **migration**, and because of the different categories of migrants, the evolution of migratory flows can only be the subject of broad estimations. On the whole, the flow of international migration has slowed down in the past years, and societal and economic constraints are likely to hinder massive migration flows.[6] Proportionately, however, immigration from developing countries to developed ones is on the rise. Roughly 2 million people per year are expected to migrate from developing to developed countries, with only 10% coming from the least developed regions.[7] In 2005, France, Germany, Italy, Spain and the United Kingdom hosted more than 29 million international migrants. Over the last few years, net migration gains for the EU ranged between 1.5 and

1.7 million per annum, including the regularisations implemented by Italy and Spain among others. That means that a considerable share of these migrants were already living in the EU. Europe (Belarus, Russia and Ukraine excluded) was the primary recipient of a stock of 44 million international migrants, on a par with the US but behind Asia (53 million).[8]

Assumptions on net immigration rates towards the EU for the foreseeable future can only be very broad, ranging between around 600,000 to 1 million per year,[9] while for the US the figure is significantly higher (1.2 million). Whereas the stock of migrant population and the pace of migratory flows towards the US will sustain considerable demographic growth and provide sufficient workforce numbers, this is unlikely to be the case in Europe. According to the United Nations, about 1.6 million migrants would be required on an annual basis, over the next decades, for ageing Europe to preserve its workforce at current levels. Given the very low fertility rate posted by most European countries, the number of migrants required to maintain the potential support ratio (i.e. the ratio of the population aged 15-64 to those aged 65+) constant would need to be much higher still.[10]

Forced migration (due to natural disasters, armed conflicts, etc.) cannot be foreseen either but can have a locally devastating impact.[11] The UNHCR has reported a constant decrease in the worldwide number of refugees, from 12 million in 2000 to 8.4 million in 2005 – the lowest level since 1980. On the other hand, since 2003, the number of internally displaced persons (IDP) has been constantly rising (from 4.3 to 6.6 million in 2005). The same goes for those considered to constitute a 'population of concern'[12] by the UNHCR (20.8 million in 2005). This increase is primarily caused by the situations in Iraq and Somalia but also in Colombia, where 2 million IDPs are recorded.[13]

In 2030, more than 60% of the world population will live in cities. **Urbanisation** ratios will be higher in developed (81.7%) than in developing countries (57%).[14] Urbanisation, however, will pose a bigger challenge for the latter, where the transition from rural to urbanised societies is likely to be uncontrolled and unbalanced.

Demographic and urban explosion, combined with environmental degradation and global warming,[15] may lead to the emergence of 'new' **diseases** (new strains of viruses or newly identified endemic viruses), the re-emergence of old ones, and the widening of areas struck by diseases that used to be regionally constrained.[16] Africa and Asia are particularly affected, the former hit by AIDS, malaria and dengue, and the latter particularly affected by tuberculosis (TB) and respiratory infections. Dengue fever is currently the most rapidly disseminating vector-borne disease, but re-emerging drug-resistant diseases, such as TB, will also represent a real threat. Worldwide, cases of TB have grown by 20% over the last ten years. 1.7 to 2 million people died from it in 2004, and 8.8 million are infected each year. Some forecast that 35 million people could die from TB by 2025, should this trend persist.

Malaria and HIV/AIDS, which are likely to remain the deadliest diseases,[17] combined with TB, kill 6 million people per year. Those diseases, but also dengue, fevers, diarrhoeal and respiratory infections, are hindering the economic development of entire regions. The WHO estimates the loss of economic growth in malaria's endemic areas of Sub-Saharan Africa to be 1.3%, and estimates the loss of GDP caused by TB to be between 4% and 7%.[18] Prospects of reversing these trends are uncertain and depend as much on future vaccines and prophylactic treatments as on the social and economic development of the areas concerned. Global warming could also provoke migration of vector-borne species carrying infectious pathogens but the scale and impact of such migration cannot be anticipated. New pathogens represent yet another kind of risk. One or two of them are iden-

Population growth in major regions and countries by age group

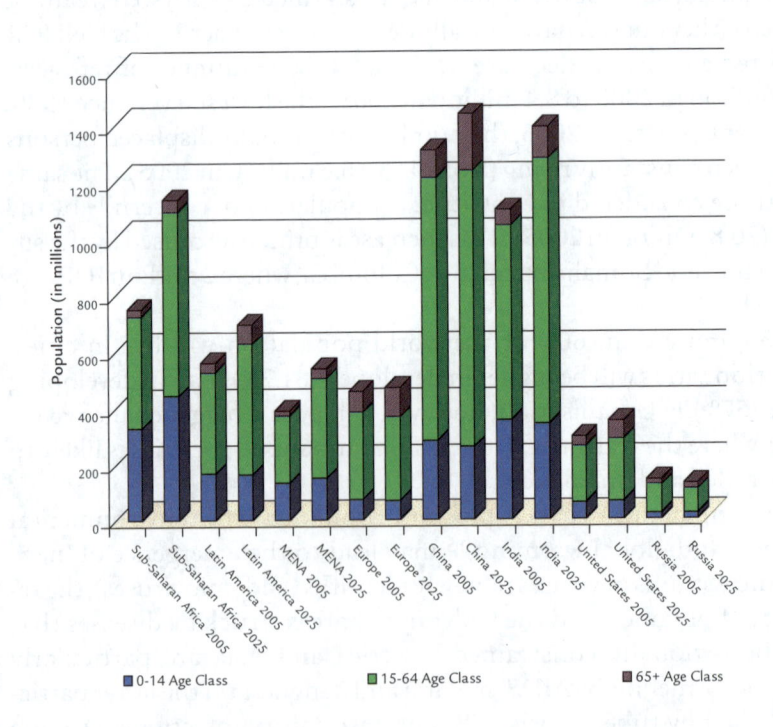

Source: United Nations Department of Economic and Social Affairs, Population Division. Data accessible at http://esa.un.org/unpp.

tified each year and their potential mutations are still unpredictable (e.g. avian influenza and SARS).[19] These pathogens, as well as drug-resistant strains, can imply epidemiological risks that may have direct economic impacts on agriculture, trade and migration.

Regional differences

Developed world

Most developed countries will see their population growth rates fall (indeed some will even turn negative) and their elderly populations expand. The median age[20] in these regions is expected to rise from 38.6 today to 43.1 in 2025 (worldwide from 28.1 to 32.8). However, important differences exist, in particular because of varying fertility rates. The countries particularly concerned by population shrinking and ageing will need to compete for skilled workforce from abroad.

The European Union (EU-25)

Based on the interplay of immigration and population ageing, the population of the 25 current EU Member States is expected to grow from 458 million in 2005 to 470 million in 2025, before decreasing.[21] Demographic trends will not be uniform (steady growth in France and the UK until 2025, early decline in Germany, Italy and Spain, among others). In all Member States, however, low fertility rates (EU average 1.5) and long life expectancy will lead to population ageing (+37.4% of people aged 65-79 years, with an EU-wide median age of 44.5, rising to even 50.5 in Italy). It is noticeable that, since the end of the 1980s, demographic growth in the EU has resulted more from immigration than from natural increase.[22]

Up until 2025, taking into account the contribution of migrants, higher participation rates of older people and women will probably be sufficient to maintain a sizeable workforce, but the recruitment of qualified and highly skilled labour is likely to prove increasingly difficult. After 2030, the working age population will begin to shrink. Twenty years from now, the effective economic old-age dependency ratio will rise sharply from 37% to 48%, with potentially huge consequences for pension schemes.[23]

Controlled immigration and incentives for qualified foreign human capital will likely become increasingly important to relieve the strained European labour markets. Immigration to the EU is expected to vary between 600,000 and 1 million per year, mainly directed to Western Europe and, increasingly so, to Southern countries such as Italy and Spain, whereas

Eastern Europe is projected to lose around 75,000 migrants per year.[24] It is generally assumed that this influx can only mitigate the economic effects of population shrinking and ageing in Europe.

The United States

The US population is expected to grow from 296.4 million in 2005 to more than 364 million in 2030.[25] This increase is due to fertility rates close to the replacement rate (2.1 children per woman), but even more so to longer life expectancy and immigration. The share of the working age population (defined here as 20-64 years) will decrease slightly from 59% to 55.2%, but immigration and higher workforce participation rates (older people, women) are expected to be sufficient to compensate for this trend.

Racial and ethnic distribution will evolve considerably. By 2030, the share of the white population (Hispanics excluded) is expected to decrease to 57.5% (from 69.4% in 2000). The proportion of African-Americans will remain stable (13.9%), whereas the Asian and Hispanic populations are expected to more than double their shares to 6.2% and 20.1% of the total population.

Immigration – both legal and illegal – has traditionally been a major element of US population growth. According to UN estimates, net migration to the US will average 1.1 million per year over the next decades. In 2004, an estimated 800,000 people entered the US illegally (57% from Mexico), driving the number of undocumented migrants up to 10.3 million.[26] Immigration pressure, in particular from Mexico and other Latin American countries, will certainly continue, with significant implications for US domestic politics and culture. However, security concerns may lead to stricter immigration policies and reduce both legal and illegal migration.

Japan

Japan is likely to suffer a demographic collapse as the century progresses. The total population is expected to decline from 128 million today to 124.8 million in 2025, while the median age will increase from 42.9 to 50. Population ageing will increase old-age dependency ratios dramatically and lead to a net decline in the labour force.[27] High cultural barriers will make it difficult for immigration to ease these developments.[28] A large number of the legal immigrants (343,000 in 2002) are specialised in show business (123,000), while engineers and specialised workers only represent a fraction of this total (2,800 and 1,800 respectively). In 2002, undocumented migrants in Japan amounted to only 250,000 people, compared to 1.8 million foreigners legally residing in the country.[29]

Russia

Like the majority of former Soviet Union Republics, Russia is facing a steep decline in its population. Population growth rates have been negative since the mid-1990s (-0.02% in 1995) and are expected to fall further to -0.58% on an annual basis between 2020 and 2025. Despite a slight migration inflow and a marginally rising fertility rate (from 1.40 to 1.58), Russia's population is expected to fall from 143.2 million to 129.2 million people over the next twenty years.[30] The increase of the proportion of elderly people, aged over 60 (from 17.1% to 24.3% of the population by 2025) could result in serious social problems, notably given the corresponding stagnation of the percentage of young people, at around 16% of the population. At the same time, the country has one of the highest rates of HIV infection in the world, with severe social and economic consequences.[31] Certain worst-case scenarios foresee 11 million HIV infections and 8.7 million HIV/AIDS-related deaths between today and 2025.[32]

Developing world

In many developing countries, birth rates and population growth rates are expected to remain relatively high over the next few decades. The population of developing countries will remain relatively young (with a median age increase from 25.6 in 2005 to 30.8 in 2025) and their working-age population will expand. The difference in demographic patterns is shifting the global distribution of population increasingly from the developed to the developing world.

The number of people living in cities in the developing world is expected to almost double from 2.2 billion today to 3.9 billion people in 2030. 15 of 22 mega-cities (cities with populations of over 10 million inhabitants) will be located in developing countries. In many regions, in particular in Africa and in India, rural economies will thus coexist with urban centres with high population concentrations.

Emigration from developing to developed countries will continue, but is unlikely to increase dramatically. Asia is expected to suffer the biggest net outflow (1.2 million per year), followed by Latin America/the Caribbean (570,000) and Africa (330,000).[33] However, notably in the case of small countries, these migration rates may entail a significant 'brain drain' and the loss of the most innovative part of the workforce, which, in turn, may hinder economic development.[34]

Sub-Saharan Africa

Due to high fertility rates (from 4.7 in 2005 to 3.5 in 2025), population in Sub-Saharan Africa is forecast to grow from 731.5 million today to 1.04 billion in 2025 (+43%). Nonetheless, with the noticeable exception of Nigeria, which will account for 190 million people (131 million in 2005), no demographic power will emerge. Life expectancy should rise from 45.8 years in 2005 to 52.3 in 2025,[35] but could be considerably lower in the 'AIDS Belt' of Southern Africa.[36] The proportion of people belonging to the younger generations will remain high (59% under 24 years by 2025) and the median age low (21.8 years, North Africa included). Very different trends can be detected at the regional level, with population growth nearly equal to zero in the southern part of the continent (+0.2 %, 55 million people in 2025), but high in Western Africa (+52%, 401 million people) and in Eastern Africa (+55%, 447 million) and very high in Central Africa (+68%, 184 million people).[37]

However, Sub-Saharan Africa's demographic trends depend heavily on future progress in the fight against diseases. In particular, the development of vaccines for malaria and AIDS could affect demographics considerably. Today, 9% of Africa's adult population (29.4 million) is HIV-infected.[38] Accounting only for 11.4% of the world's population, the continent harbours 70% of all HIV infections and accounts for 77% of HIV-related deaths.[39] Between 67 and 83 million casualties are expected from 1980 to 2025. In parallel, an increase in cases of malaria and in cases of dengue fever is expected.[40] Malaria kills 900,000 Africans per year (90% of the world total) and tuberculosis killed 587,000 in 2004. The spread of other resilient diseases (cholera, tuberculosis) and the negative impact of human pollution could increase death rates even more.

In 2025, Africa will still be an empty continent, with relatively low population density and urbanisation (50% in rural areas). The degree of urbanisation will be lower in Africa than in Asia and Europe, but the annual growth rate of urbanisation is projected to be higher (Lagos +4.51%, Kinshasa +4.03%, etc.).[41] Around 2020, Africa will have 11 major cities with a population of five million or more and over 700 cities with a population surpassing 100,000.[42] The impact of demographic growth and rural migration may be severe, in particular because of inadequate infrastructures and competition for resources.

The Middle East and North Africa

Fertility rates are expected to continue to fall from 2.91 to 2.38 in North Africa and from 3.11 to 2.52 in the Middle East, slowing down population growth rates in the MENA region as a whole to 1.3% in 2025 (from 3.4% in 1985 and 2% today). By 2025, however, the population is expected to grow by 38% (from 388 to 537 million).[43]

The working-age population will expand particularly fast (+40% in North Africa, almost +50% in the Middle East).[44] Given the social and economic prospects in many MENA countries, part of this workforce will probably emigrate. If current trends continue, Europe is likely to remain the main destination for emigration from North Africa. This dynamic might be enhanced if the traditional recipient countries in the region – those belonging to the Gulf Cooperation Council (GCC) – continue to restrict the influx of foreign migrants in order to relieve their labour markets and reserve jobs for the local population. The presence of foreigners in the workforce is close to 90% in the United Arab Emirates (UAE) and Qatar, and 60% in Saudi Arabia and Bahrain. Foreigners represent 38.5% of the GCC countries' global population, with shares ranging from 26% to 80% and Arabs representing less than one third and outnumbered by Pakistani and Indian workers. The GCC countries have already started to massively expel their foreign workforces, in particular those coming from India, Pakistan, Egypt and Yemen.[45] This, in turn, cuts off the population in the migrants' countries of origin from an important source of revenue in the form of remittances.

Asia

Asia will remain the most populated continent (3.9 billion people today, 4.7 billion in 2025) with a still relatively young population (median age 27.7 today, 33.7 in 2025). However, trends differ considerably between the various countries: Pakistan's population, for example, will increase from 158 to 229 million, and its fertility rate of around 2.6 will still stand well above the replacement rate. Indonesia's population will increase from 222 million to 267 million, but its fertility rate will drop to 1.85.[46] Yet the whole area will be under heavy demographic pressure, which will impact the most on those countries posting feeble rates of economic growth, such as Pakistan and Bangladesh (186 million people in 2025).

Migration within and out of Asia is expected to remain substantial and play an important role for the economy of the least developed countries of the continent (with local populations depending on remittances).[47]

The major countries of net emigration in Asia are expected to be China (-327,000 annually), India (-241,000), the Philippines (-180,000), Indonesia (-164,000) and Pakistan (-154,000).[48] The GCC, US, Japan, Australia and some relatively well-off states of South East Asia are the main destinations of those fluxes.

Since the emergence of new viral diseases seems to be increasingly linked to the proximity between human beings and animal species, the high density of animal and human populations cohabiting in Asian cities could represent an increasing threat for human health. Moreover, Asia, and South Asia in particular, is exposed to a steep increase in HIV contamination.[49]

China

By 2025, the Chinese population is forecast to grow from 1.31 to 1.44 billion. By then, the proportion of the working-age population will still be high (68.4% of the population), but two decades of the one-child-policy will increasingly make its impact felt on the population's age structure.[50] The median age in China will increase from 32.6 today to 39.5 in 2025 (and 43.0 in 2035).

From 2015 on, population ageing will lead to a rapidly increasing number of elderly people (from 7% of the total population today to 20% in 2025) and to a steadily increasing dependency ratio (40.2 in 2015, 50.5 in 2030). In 2015, the young-age dependency ratio will be 26 and the old-age dependency ratio 13; in 2030 these figures will be around 25 for young-age dependency and 24 for old-age dependency, with a marked progression of the latter and a consequent increment in the social costs that it will entail.[51] The rapidity of this shift is likely to create considerable social problems and to at least slow down China's economic growth.

Increasing discrepancies in living standards between rural areas and cities are expected to provoke migration towards urban areas, in particular in the coastal regions. Consequently, the urbanisation rate is expected to rise from 41% today to 57% in 2025 (824 million people living in cities), with 85% of the population living in the Central and Eastern regions of the country.[52] This means that a huge imbalance is emerging between the extension of the (mostly empty) national territory and urban density.[53]

India

India's demographic development will remain dynamic. Fertility rates will remain above the replacement level (2.76 in 2005, 2.11 in 2025) and life expectancy will grow (from 65 years in 2005 to 70 in 2025). India's total population will come close to China's by 2025 (1.39 versus 1.44 billion), but will remain considerably younger (the median age will be 30.4 in 2025 and 33.9 in 2035). The old-age dependency ratio, with an elderly population (people aged over 65) doubling to about 80 million by 2025 (only 6% of the population) will remain very low as well.

Demographic distribution across the country will remain highly uneven. In 2025, India will still be a mainly rural country with 62% of its population living in rural areas (71.3% today). Population density will nevertheless be very high (425 inh/km^2 in 2025, as opposed to 336 inh/km^2 today). Moreover, population growth, compounded by internal migrations, will notably impact on 60 or 70 large urban centres with over 1 million people, in a country where only 30% of the population has access to improved sanitation and urban structures are already under huge strain. Some experts consider that HIV/AIDS could turn into an epidemic in the coming years. Up to 25 million AIDS cases could occur in India by 2010. Cumulative AIDS deaths from 2000 to 2015 could reach 12.3 million and 49.5 million from this date up to 2050.[54]

Notes

1. *World Population Ageing: 1950-2050* (Department of Economic and Social Affairs, Population Division, United Nations, New York, 2002).
2. Inevitably, there are noticeable differences between different projections. See, for instance, Wolfgang Lutz, James W. Vaupel and Dennis A. Ahlburg (eds.), 'Frontiers of Population Forecasting', *A Supplement to the Population and Development Review*, vol. 24, 1998. The figures reported in the text are extracted from United Nations, World Bank, US Census and EU statistics, and are based on medium projections. Relevant data can be consulted online at http://esa.un.org/ unpp, http://devdata.worldbank.org/hnpstats/dp.asp, http://epp. eurostat.cec.eu.int, http://www.census.gov and in *World Population Prospects: The 2004 Revision* (United Nations, New York, 2005).
3. *World Population Ageing: 1950-2050*, op. cit.
4. *Trends in International Migration* (SOPEMI: OECD, 2004).
5. *World Population Ageing: 1950-2050*, op. cit.
6. *Trends in Total Migrant Stock: The 2005 Revision* (Department of Economic and Social Affairs, Population Division, United Nations, February 2006); *World Population Monitoring, Focusing on International Migration and Development* (Commission on Population and Development, Economic and Social Council, 25 January 2006) and *World Economic and Social Survey: International Migration* (Department of Economic and Social Affairs of the United Nations, 2004).
7. For the UN definition of least developed regions, and the list of the 50 countries included in this cluster, see the United Nations Department of Economic and Social Affairs, Population Division: http://esa.un.org/ unpp.
8. Online data from the United Nations Department of Economic and Social Affairs, Population Division (http://esa.un.org/migration/ index.asp?panel=1) and *Trends in Total Migrant Stock: The 2005 Revision*, op. cit. It should be noted that the calculation of migrant flows and migrant stocks largely depends on respective definitions.
9. See: *The impact of ageing on public expenditure: projections for the EU25 Member States on pensions, health care, long-term care, education and unemployment transfers (2004-2050)*. The Economic Policy Committee and the European Commission (DG ECFIN), *Special Report* no. 1/2006, and Tobias Just and Magdalena Korb, 'International Migration: Who, Where and Why?' in *Current Issues: Demography Special* (Deutsche Bank Research, Frankfurt, 1 August 2003).
10. *World Economic and Social Survey: International Migration*, op. cit.
11. Palestinians (4 million people) are not counted. *Global Refugee Trends* (Population and Geographical Data Section, Division of Operational Support, United Nations High Commissariat for Refugees, 17 June 2005).
12. Population of concern includes, according to the definition of the UNHCR: 'Refugees, asylum-seekers, returnees (refugees who have returned during 2005), internally displaced persons (IDPs), returned IDPs (IDPs who have returned to their place of habitual residence during 2005), stateless persons, and others of concern not falling under any of the categories above'; *2005 Global Refugee Trends* (UNHCR, 9 June 2006).
13. *2005 Global Refugee Trends*, op. cit.
14. *World Urbanization Prospects: The 2003 Revision* (Department of Economic and Social Affairs, Population Division, United Nations, 2003).
15. *Global Environment Outlook Year Book: An Overview of our Changing Environment 2004/5*, United Nations Environment Programme (UNEP), 2005.

16. *The Global Infectious Disease Threat and its Implications for the United States* (National Intelligence Estimates, NIE 99-17D, January 2000); Erik R. Peterson, 'SARS, Lessons for the Longer Term', CSIS (Center for Strategic and International Studies), *FYI 9*, April 2003.
17. AIDS has killed more than 25 million people since 1981. Currently, 40.3 million people (average estimate) are infected, and 3.1 million people (average estimate) died in 2005. US figures foresee about 100 million HIV cases in 2010. According to the World Health Organization, malaria is killing roughly one million people per year. *AIDS Epidemic Update* (UNAID/World Health Organization, 2005); *Addressing the HIV/AIDS Pandemic: A US Global AIDS Strategy for the Long Term* (Council on Foreign Relations and Milbank Memorial Fund, 2004); *2004 Report on the Global HIV/AIDS Epidemic: 4th Global Report* (UNAIDS, Geneva, 2004). L. Gallup and J. D. Sachs, *The Economic Burden of Malaria*, CDI Working Paper 52, Center for International Development, Harvard University, Boston, 2000.
18. See WHO internet portal on malaria and TB, and *HIV/AIDS, Tuberculosis and Malaria: the Status and Impact of the Three Diseases* (The Global Fund to Fight Aids, Tuberculosis and Malaria, 2005).
19. Paul Rincon, 'Faster emergence for diseases', BBC News, 20 February 2006. According to the WHO, 'between 1972 and 1999, 35 new agents of disease were discovered and many more have re-emerged after long periods of inactivity, or are expanding into areas where they have not previously been reported'. *Emerging Issues in Water and Infectious Diseases* (World Health Organization, 2003).
20. The median age is the 'Age that divides the population in two parts of equal size, that is, there are as many persons with ages above the median as there are with ages below the median' (UN definition).
21. 'Confronting Demographic Change: a new solidarity between the generations', European Commission, Green Paper, COM (2005) 94 Final, 16 March 2005; Eurostat (http://epp.eurostat.cec.eu.int); Klaus Regling, 'How Ageing Will Torpedo Europe's Growth Potential', *Europe's World*, Spring 2006; Jonathan Grant et al., *Low Fertility and Population Ageing: Causes, Consequences, and Policy Options*, MG-206-EC (Rand Corporation, Santa Monica, California, 2004).
22. Alain Monnier, 'Les évolutions démographiques de l'Union européenne', *Questions internationales*, no. 18, La Documentation française, Paris, March/April 2006.
23. The effective economic old-age dependency ratio is defined as the non-active population aged 65+ as a percentage of the employed population aged 15-64. See *The impact of ageing on public expenditure*, op. cit.
24. Tobias Just and Magdalena Korb, 'International Migration: Who, Where and Why?', op. cit.; *Replacement Migration: is it a Solution to Declining and Ageing Populations?* (United Nations Population Division, 21 March 2000); Lindsey Grant, 'Replacement Migration: the United Nations Population Division on European Population Decline', *Population and Environment*, vol. 22, no. 4, 2001; D. A. Coleman, *Who's Afraid of Support Ratios? A UK Response to the UN Population Division Report on 'Replacement Migration'*, Paper prepared for United Nations 'Expert Group' meeting held in New York, October 2000, to discuss the report on 'Replacement Migration' presented by the United Nations Population Division, Shorter Revised Draft, 20 December 2000.
25. US Census Bureau, 2004, 'US Interim Projections by Age, Sex, Race, and Hispanic Origin', http://www.census.gov/ipc/www/usinterimproj/; internet release date: 18 March 2004.
26. See: Jeffrey S. Passel, 'Estimates of the Size and Characteristics of the Undocumented Population' (Pew Hispanic Center, Washington D.C., 2005).
27. For a more detailed economic approach, see for instance, Masakazu Inada, Takashi

Kozu, Yoshiko Sato, 'Demographic Changes in Japan and their Macroeconomic Effects', Report no.04-E-6, Bank of Japan, September 2003; Hamid Faruqee and Martin Mühleisen, 'Japan: Population Ageing and the Fiscal Challenge', *Finance and Development* (IMF), vol. 38, no. 1, March 2001.

28. From 2000 to 2002 (last data available), foreign immigration in Japan stood at around 345,000 people per year (Chinese and Filipinos representing roughly half of the migrants). Those high figures do not only reflect the attractiveness of the Japanese economy: every person who stays more than 90 days on the territory is considered as a migrant. See *Trends in International Migration...*, op. cit.

29. *Trends in International Migration...*, op. cit.

30. United Nations Department of Economic and Social Affairs, Population Division: http://esa.un.org/unpp.

31. According to some scenarios, 640,000 to 1.2 million people could be infected by 2008. 'The Economic Consequences of HIV in Russia', The World Bank, Russia Office, 10 November 2002; S. Misikhina, V. Pokrovsky, N. Mashkilleyson, D. Pomazkin, 'A model of social policy costs of HIV/AIDS in the Russian Federation', Research and Policy Analysis, International Labour Office, Geneva (undated).

32. Celeste A. Wallander, *The Impending AIDS Crisis in Russia: The Shape of the Problem and Possible Solutions*, Conference at the CSIS HIV/AIDS Task Force, Center for Strategic and International Studies, 14 April 2004 (http://www.csis.org/media/csis/events/040415_wallander.pdf).

33. *World Population Prospects: The 2004 Revision*, op. cit.

34. It is estimated that Africa, for example, has lost one third of its qualified human capital and that the rate of loss is increasing. Since 1990, on average 20,000 individuals with high levels of training and education (such as doctors, engineers, and university lecturers) have left the continent each year. Overall, there are about 300,000 highly qualified Africans abroad, of whom 10% hold Ph.Ds. This data is based on information from the International Organization for Migration (IOM) quoted in 'African Regional Document: Water Resources Development in Africa', 4th World Water Forum, 16-22 March 2006. Accessible at http://www.worldwaterforum4.org. mx/home/tools.asp?lan.

35. World Bank statistics (http://devdata.worldbank.org/hnpstats/ dp.asp) and United Nations Population Division statistics (http:// esa.un.org/unpp/index.asp?panel=2). According to the UN, because of HIV/AIDS, a deficit of 84.1 million people with respect to potential growth rates is foreseen for the forty most affected African countries in 2015. *World Population Prospects: The 2004 Revision*, op. cit.

36. *2004 Report on the Global HIV/AIDS Epidemic: 4th Global Report* (UNAIDS, 2004).

37. United Nations figures, accessible at http://esa.un.org/unpp/index.asp?panel=1.

38. *AIDS in Africa: Three Scenarios to 2025*, Joint United Nations Programme on HIV/AIDS (UNAIDS, 2005); *World Population Prospects: The 2004 Revision*, op. cit.

39. 'Community Realities & Responses to HIV/AIDS in Sub-Saharan Africa' (United Nations, New York, June 2003).

40. *Millennium Ecosystem Assessment. Ecosystems and Human Well-Being: a Health Synthesis* (World Health Organization, 2005); *Climate Change Futures: Health, Ecological and Economic Dimensions* (The Center of Health and the Global Environment, Harvard Medical School, November 2005).

41. Out of sixteen cities, seven African cities will be confronted with the world's highest rates of growth from 2000 to 2015. *World Urbanization Prospects: The 2003 Revision*, op. cit.

42. See: 'African Regional Document: Water Resources Development in Africa', op. cit.

43. These figures include Turkey and Iran in MENA. See the United Nations Department of Economic and Social Affairs, Population Division: http://esa.un.org/unpp.
44. http://devdata.worldbank.org/hnpstats/dp.asp; *Unlocking the Employment Potential in the Middle East and North Africa: Toward a New Social Contract* (MENA Development Report, The World Bank, Washington, D.C., 2004).
45. See: Martin Baldwin-Edwards, *Migration in the Middle East and Mediterranean* (Mediterranean Migration Observatory, University Research Institute for Urban Environment and Human Resources, Panteion University, Greece, January 2005).
46. United Nations statistics: http://esa.un.org/unpp.
47. *Levels and Trends of International Migration to Selected Countries in Asia* (Department of Economic and Social Affairs, Population Division, United Nations, 2003).
48. *World Population Prospects: The 2004 Revision*, op. cit.
49. *AIDS in Asia: Face the Facts. A Comprehensive Analysis of the AIDS Epidemics in Asia* (Monitoring the AIDS Pandemic [MAP] Network, 2004); *The Next Wave of HIV/AIDS: Nigeria, Ethiopia, India, Russia and China* (National Intelligence Council, Washington, D.C., Sept. 2002).
50. See for instance Wolfgang Lutz et al., 'China's Uncertain Demographic Present and Future', Interim Report IR-05-043 (International Institute for Applied Systems Analysis, Austria, 5 September 2005).
51. UN figures, accessible at http://esa.un.org/unpp/p2k0data.asp
52. Tian Xiang Yue et al., 'Surface modelling of human population distribution in China', *Ecological Modelling*, no. 181, 2005.
53. In 2000, 42.1% of the 667 major cities of China were located in Eastern China, whose area accounts for 9.5% of the country, 34% of these cities were distributed in Middle China (17.4% of the country's area), and only 23.8% of them in Western China (70.4% of the country's area). The density of cities in Eastern China and in Middle China was, respectively, 13.1 times and 5.8 times higher than the density in Western China. In other words, large swathes of the country are relatively empty, and the impact of urbanisation needs to be assessed on a regional basis. See Tian Xiang Yue, 'Surface modelling of human population distribution in China', op. cit.
54. *HIV/AIDS. Country Profile: India* (UNAIDS, Population Division of the Department of Economic and Social Affairs of the United Nations Secretariat, March 2003); *World Population Prospects: The 2002 Revision. Highlights* (United Nations, New York, February 2002); *The Next Wave of HIV/AIDS: Nigeria, Ethiopia, Russia, India and China*, op. cit.

Economy

Economic development is interlinked with other key variables such as politics, security, demography, technology, and energy. Economic performance can be measured by various indicators such as productivity, investment, trade, and financial flows. All this makes future economic trends particularly difficult to predict. However, there is a general consensus on some important broad trends.

> ### General Trends
>
> - Economic globalisation will continue and deepen. In spite of recurrent protectionist tendencies, cross-border trade is expected to grow worldwide.
>
> - Greater openness and new trade powers will enhance competition and increase the pressure to adapt economic and social systems, which will pose different challenges to developed and developing countries. Countries and regions that fail to adjust will risk marginalisation.
>
> - The dominant triad – the US, the EU and Japan – will probably remain leaders in many high-value markets, but continue to offshore labour-intensive production and business services.
>
> - Some of the target countries for these offshore activities will become new economic powerhouses. China and India, especially, are likely to continue their economic rise, shifting the centre of the world economy to Asia. However, in both countries, sustainable economic growth will depend on domestic reforms, energy supply and the development of infrastructures.

2 The New Global Puzzle — Part I

> ▶ Emerging economies in Asia are expected to post the highest growth rates, with Chinese and Indian GDP expected to triple by 2025. However, in terms of GDP per capita, they will remain a long way behind OECD countries. The distribution of wealth will thus remain highly unequal.

Over the next twenty years, **economic globalisation** will most probably continue to drive economic growth. World growth is expected to remain sustained, with an average annual rate of 3.5% between 2006 and 2020. By then, world GDP is expected to be two thirds bigger than in 2005 ($44.4 trillion).[1] Provided unforeseen major events do not occur, international trade, investment and capital flows are expected to steadily grow worldwide, leading to further integration of markets for goods and services. At the same time, these flows will continue to diversify geographically towards emerging markets. Globalisation will therefore entail momentous, and sometimes contradictory, consequences. Greater openness and growing competition will impose structural adjustments on all countries and will put not only economic but also social models under stress, which could be painful at the domestic level and lead to tensions on the international stage.

New competitors have emerged on the world market, challenging developed and developing economies across a broadening range of industries. Thanks to rising levels of education and investment, China and India are beginning to diversify their economies towards sectors with higher technological added value.[2] On the whole, **technological progress** will play a key role in fostering globalisation. In technology-intensive industries, innovation is associated with radical changes in production methods and organisation (flatter structures, outsourcing and transnational alliances).[3] Companies will increasingly be able to outsource any given segment of production to the most cost-effective location.[4]

Service offshoring is still at an early stage but is growing very fast, with an envisaged expansion of offshore outsourcing of business processes from $1.3 billion in 2002 to $24 billion in 2007.[5] New information and communication technologies have increased the tradability of information-centred sets of services. Firms will increasingly profit from offshoring these activities to a foreign country, either through FDI or outsourcing.[6]

In parallel to these intensifying trends, **FDI and trade flows** will expand and take new directions. Most notably, developed countries will no longer be the only large investors. Over the last 15 years, FDI outflows from developing countries have grown faster than those from developed countries (although from a negligible level in the early 1990s). They accounted for one tenth of the global FDI stock ($0.9 trillion) in 2003. In particular, FDI among developing countries is growing faster than investment from developing to developed countries.[7] In 2000, it represented more than 30% of the FDI inflows reported by developing countries.[8]

The diversification of trade flows has already taken place on a much larger scale, and will most likely continue at an accelerating pace. The consolidation of integrated regional economic areas represents one of the main trends of international trade. In particular, **East Asia is emerging as a key regional trade area**. Between 1984 and 2004, developing Asia's share of world exports almost doubled and reached 21.3%. The share of intra-regional trade flows has increased from 22% (1984) to about 40% (2004) of developing Asia's total exports (54% if Japan is included), compared to 46% in North America and 64% in the EU.[9] Trade routes follow a new regional and international division of labour. Chinese exports of final goods to the US and EU are expanding rapidly. Other industrialised Asian countries have lost shares on those markets, but are increasing their exports of intermediate and capital goods to China.[10]

Asia still depends on Western markets, and notably on US domestic demand, for the export of final goods. However, the growing demand from the fast expanding **middle-classes in emerging Asian countries** will probably determine considerable shifts in trade patterns, and create new export opportunities for developed economies. The middle class (those with an income higher than $3,000 per year) in the four largest emerging economies (China, India, Brazil and also Russia, whose economy seems on the way to recovery) could be 800 million-strong in less than a decade. If high growth expectations are fulfilled, 200 million people with an income above $15,000 could be added to these economies by 2025.[11] Rapid income rise will result in a growing domestic market. For example, Asia is likely to become the world leading automotive market as soon as the next decade.[12]

Another important feature of global exchanges is the **proliferation of preferential trade agreements, especially bilateral free trade agreements**.[13] Compared to multilateral liberalisation at the global level, advocated by the WTO in the Doha Round, preferential agreements may address a wider range of issues including investment, competition, environment and labour standards, and lead to deeper economic integration among those involved. However, the complex network of overlapping regional and bilateral agreements may lead to conflicts of regulatory standards and undermine transparency and predictability. Failure to spur multilateral trade liberalisation during the Doha Round will likely foster more preferential agreements. Ensuing fragmentation would provide lower global gains and mainly benefit key economic and trade 'hubs' (such as the EU and the US) compared to other countries.

The globalisation of exchanges and investments benefits a growing range of countries, but not all. Those countries left out of global trade and investment flows risk further marginalisation. At the global level, **income inequalities** between countries are narrowing. Progress in this direction, however, largely derives from the economic growth of populous South and East Asian countries, especially China and India. On the other hand, the gap between developed (and emerging) countries and Sub-Saharan Africa has been widening. With a view to offsetting this trend, countries in Sub-Saharan Africa, as well as other least developed countries, would need to achieve and maintain growth rates well above the world average.

Likewise, while extreme **poverty** has been declining globally, this has not been the case in every region.[14] Based on the World Bank definition of extreme poverty – those living with less than $1 per day – the numbers have gone down. 400 million fewer people lived with less than $1 per day in 2001 than in 1981. High growth rates in Asia have driven the poverty reduction process, especially in countries like China, India and Bangladesh.[15] In contrast, Africa has experienced an increase in the number of people living in absolute poverty, from 160 million in 1981 to 303 million today. If the current levels of economic growth prove sustainable, the Millennium Development Goals' target of halving the number of people living with less than $1 a day by 2015 (from 1,218 million in 1990 to an expected 617 million) will be met, but only because progress in Asia will offset the opposite trend in Sub-Saharan Africa. In this region, those in dire need are expected to increase to 336 million (or 38% of the population in 2015).[16]

Taking yet another international standard of reference to measure poverty – those living with less than $2 per day – poverty reduction proves slower. The number of people living on less than $2 per day has only slightly decreased since 1990, from 2.65 billion to 2.61 billion, despite the rapid drop in China (from 825 to 533 million).[17] The envisaged drop from 2.61 to 1.99 billion by 2015 will largely stem from economic development in the Asia-Pacific region, and especially in China.[18] When it comes to poverty reduction, therefore, the pace of progress will vary greatly depending on the region at stake and on the parameter of reference.

Clearly, considerable **disparities** will therefore persist between different countries, which will be exposed to different challenges in the process of globalisation. A key distinction should be drawn between developed and developing countries. **Developed countries** will need to confront a process of economic adaptation to growing competition, and transfer their resources from declining activities to sectors where they enjoy a comparative advantage. This evolution, together with the offshoring of a (limited) part of their industrial base and of business services, and increasing international competition, will probably lead to considerable social tensions, potentially leading to protectionist reflexes.[19]

Developing countries will be confronted first and foremost with the tasks of building adequate physical infrastructures, enhancing their human capital and improving governance structures and the investment climate. The least developed countries may face yet another serious obstacle to their development. Although their economic growth will at least partially depend on trade liberalisation, the consequent loss of revenue from import duties, and preference erosion to the advantage of emerging trade powers, may severely affect them.[20]

As a result of ongoing trends and of the momentous changes that they will determine, rich developed countries will see their share of global wealth, trade and FDI decline in relative terms, to the advantage of emerging countries. At the horizon 2025 and beyond, **Asia will likely become one of the pivotal centres of the world economy**. The share of world GDP produced by OECD countries will shrink, in PPP terms, from around 55% in 2000 to 40% in 2030, while emerging Asia's share will rise from 24% to 38% over the same period.[21] According to some projections, the aggregated

Projected GDP growth (in US$ billions)

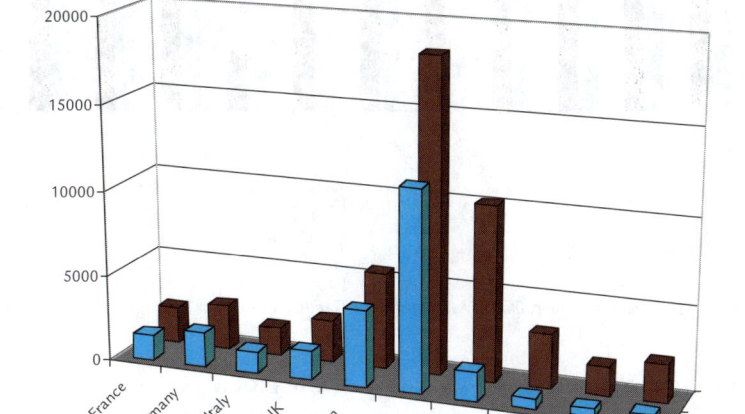

Source: Dominic Wilson and Roopa Purushothaman, *Dreaming with BRICS: The Path to 2050*, Global Economics Paper no. 99, Goldman Sachs, 1 October 2003.

GDP of Brazil, Russia, China and India could amount to half the GDP of the current six biggest developed countries (US, Japan, Germany, UK, France and Italy) by 2025 and overtake it by 2040.[22] Looking at the size of national GDP, the top five countries in 2025 will be the US, China (due to overtake Japan in 2016), Japan, India and Germany.[23]

The picture will, however, look radically different when focusing on the **levels of GDP per capita**. From a current GDP per capita of $6,300 in PPP[24] (15% of the US level), China could reach 24% of the US level in 2020 (higher than the level of Brazil today). India, with a GDP per capita of $3,400 in PPP in 2005 (8% of the US level), is expected to reach 12% in 2020, close to the level of Thailand today.[25] From this perspective, therefore, filling the gap between emerging economies and rich countries will take several decades.

GDP per capita in PPP (US=100)

Country	2005	2020
Japan	76	75
France	75	72
Germany	70	70
UK	78	78
Poland	30	38
South Korea	52	66
China	15	24
India	8	12
Argentina	33	38
Brazil	21	22
Mexico	24	23
Russia	26	33

Source: *Foresight 2020: Economic, Industry and Corporate Trends*, Economist Intelligence Unit, March 2006.

Following the envisaged shake-up of the global economic order, global economic governance will become increasingly complex, with China, India and others challenging the traditional leadership of the US, Japan and Europe in key international institutions (WTO, World Bank, IMF).[26] The growing complexity of decision-making at the global level may reinforce the trend towards bilateral and regional agreements. However, such fragmentation is not expected to reverse the general trend of globalisation.

Regional differences

The dominant triad

The three currently predominant economic blocs – the US, the EU and Japan – will face increasing competition from emerging markets and will continue to offshore labour-intensive activities. However, they are expected to maintain their industrial cores, in particular in capital- and research-intensive sectors, but also in sectors where transport costs relative to the value of the good being produced are high. At the same time, financial and investment flows will increasingly become a two-way street, with emerging countries buying into the industrial base and holding foreign debts of developed countries.

The United States

The US is expected to remain the world's leading economic power in 2025, continuing to post the highest level of GDP overall and per capita. Several comparative advantages will probably allow the US to maintain its current position: continuing population growth, in combination with high-quality human capital, flexible labour markets and high labour productivity; a strong culture of innovation, combining high spending on research in key sectors for the future (IT, biotechnology and nanotechnology) and the capacity to quickly translate new technologies into commercial applications (low level of regulation, a big domestic market, access to venture capital, etc.). The US is expected to remain the world leader in business and professional services, in particular in the ICT sector. Services are also expected to contribute positively to net trade, whereas goods trade is likely to remain in deficit.[27]

Between 1997 and 2004, the US has accounted for 46.6% of the increase in world aggregate demand.[28] The question is for how long the US will remain the world's economic engine. The main factors fuelling uncertainty as to the sustainability of US domestic demand are high budget and current account deficits. The current account deficit is close to $800 billion, or 6.4% of GDP (matched by increased surpluses in East Asia and oil-exporting countries), and the budget deficit grew from a surplus of 1.3% of GDP in 2000 to a deficit of 4.1% in 2005. On present trends, both in absolute terms and relative to GDP, both deficits are projected to continue to grow. The current budget deficit is in part driven by high defence expenditures. In the next few decades, significant strains on the budget will come from a growing elderly population and rapidly rising health care costs, which will cause federal spending for social security, Medicare and Medicaid to increase from 8% of GDP in 2005 to almost 15% in 2030.[29]

The US 'twin deficits' are also core factors of global financial and trade imbalances and among the main financial risks to the global economy. The US must import about $1 trillion of foreign capital per year to finance its deficit and its own FDI. This results in the huge and increasing share of foreign ownership of the US Federal debt (in particular by Japan and China). It is generally recognised that these imbalances are difficult to sustain and imply in particular the risk of a sharp depreciation of the dollar. To mitigate the risk of a disorderly adjustment, the US would have to stimulate household savings and reduce the public deficit, whereas the surplus countries would have to increase investment and encourage domestic spending.[30] In addition, a realignment of exchange rates is considered a necessary step to overcome the imbalance. Another, albeit related, problem for the US economy are high security costs (for both the federal budget and companies).[31]

The European Union

GDP in the EU is expected to grow more slowly than in the US. In many Member States demography will become a major drag on economic growth, with population ageing putting social and economic systems under considerable strain. Until 2025, higher participation rates (of older people and women), productivity gains and immigration will probably be sufficient to maintain a sizeable workforce. However, effective old age-dependency ratios are expected to rise sharply and raise complex questions regarding pensions and health care.[32]

Offshoring of production and business services will continue to thin out the European industrial base, implying further job losses in particular in labour-intensive sectors.[33] Offshoring, however, will be directed not only outside the Union but also towards the new Member States in Central and Eastern Europe, which offer business and industrial investors the attractions of low wages as well as geographic proximity and a sound and stable institutional environment. At the same time, core industries, such as telecommunications, mechanical engineering, automobiles or civil aircraft are expected to remain competitive and innovation leaders. However, relatively low research investment and overregulation may limit Europe's role in future key sectors.[34]

The EU's pattern of specialisation has already evolved. EU countries maintain a strong position in the field of high-quality products, but have accumulated a trade deficit for low-quality goods. In addition, the EU has lost ground in high-tech industries since the end of the 1990s, suffering like the US and Japan from China's competition. Technological innovations can, however, provide EU industry with new opportunities. In important sectors such as chemistry, where the EU retains a strong position, or pharmaceuticals, the technological revolution could dramatically change the face of the market.[35] The focus on technological innovation highlights the importance of the Lisbon process and goals for the future of the European economy. The ability of the EU to maintain its position in high-tech and high-quality products, intensive in R&D and skilled labour, will determine its growth prospects. At the same time, relatively low investment rates in ICT for services will need to be raised if service productivity is to grow, as has been the case in the US, and stimulate a positive economic cycle.

The Lisbon targets are, however, increasingly considered as 'desirable' rather than 'achievable'.[36] European R&D expenditure equals two thirds of the US level. In 2003, only 1.93% of EU GDP was devoted to R&D (against 2.59% in the US, 3.15% in Japan, and 1.31% in China), up 0.7% per year between 2000 and 2003. At this pace, the share would be 2.2% in 2010, far from the Lisbon target of 3%.[37] The low levels of private R&D funding explain this delay.[38] Moreover, Europe falls short of fully exploiting its potential: with 38.3% of global scientific publications, the EU leads world scientific output, but lags behind both Japan and the US in number of patents.[39]

Japan

After a decade of stagnation (1.4% average growth in the 1990s, and 0.4% between 2001 and 2003), Japan is experiencing relatively sustained growth rates (2.3% in 2005, down from 2.6% in 2004). The growth outlook for the next few decades is, however, the least optimistic among industrial countries. Japan will be even more affected than Europe by demographic decline. A shrinking workforce and consumer markets, on the one hand, and burdensome transfer payments on the other, will likely slow down real GDP growth[40] (estimations ranges from 0.7% to 1.1% per annum over 2006-2020).[41] Indeed, a persisting low fertility rate (currently at only 1.3 children per women) is expected to lead to a decline of the working age population by one-fifth over the next 25 years.[42] Policies aiming at increasing the labour force participation of older persons and women would be particularly important to mitigate this negative trend, as well as productivity-enhancing investments.

Rapid population ageing aggravates the key challenge facing Japan over the coming decades: to control its fiscal deficit and tackle its huge gross public debt (currently above 170% of GDP). The government plans to achieve a balanced budget balance in 2011 (from an estimated deficit of 4% of GDP in 2006, down from 6.7% in 2002).[43] Stabilising the public debt to GDP ratio would nevertheless require additional efforts.[44]

Japan's insertion in the world economy and its relations with its neighbours are evolving rapidly, due in particular to the emergence of China as the major trading power of the region, overtaking Japan. Japan's trade has largely shifted from the US to China, and is likely to continue to do so. While Japan was for a long time the first Chinese supplier, 2004 was a turning point as China became the first supplier of Japan. Japanese corporations have indeed created production facilities in developing Asia and especially China, leading to a new regional division of labour and the offshoring of segments of production.[45]

Among OECD countries, Japan exhibits the lowest level of inward FDI stock, import penetration and proportion of foreign workers in the labour force.[46] Policies aiming at opening up the economy are, however, under way. In this perspective, the government has outlined its ambitious target to double the stock of FDI as a share of GDP by 2010.

New powerhouses

China and India are expected to continue their economic growth path over the next two decades. First, they will further develop their respective roles as 'extended workbench' and 'service provider' of the industrialised world. Second, they will probably be able to further diversify their economies, with India expanding its manufacturing base and China shifting its production to more knowledge-intensive goods, and to services. Third, as illustrated above, these two countries will likely diversify their current trade patterns towards emergent demand within Asia.[47] Both countries will probably catch up considerably with the developed world, but remain in many ways developing countries (in terms of per capita GDP, social security systems, etc.). At the same time, they will need to undertake enormous investment in their energy and transport infrastructures to sustain current growth rates.

China

Economic growth in China is likely to continue over the next couple of decades, although at a slower rate. Its GDP is expected to triple by 2025, ranking second behind the US (or even overtaking it in PPP terms).[48] GDP per capita, in contrast, will still be much lower (close to a quarter of the US level in PPP in 2020).[49]

GDP growth in China will continue to be driven primarily by industrial production. This includes a gradual shift from cheap and simple items to higher-quality products. For most sophisticated capital goods and knowledge-intensive products, however, Chinese firms will probably still find it difficult to keep up with foreign competitors.[50] This does not however rule out the possibility that China could catch up partly in certain strategic areas, such as IT, biotechnology and aerospace, which Beijing has defined as research priorities.[51]

In 2025, China will be the world's largest exporter and importer. Chinese trade is expanding at an amazing pace (+35% a year between 2003 and 2005). Its exports have been particularly dynamic towards countries such as the US (+30% in 2005), and European countries such as Germany and the Netherlands (+37% and +40% respectively). Two-thirds of China's imports originate in Asia, and half of its exports go to that region. China could thus become the heart of a booming and increasingly integrated Asian region.[52] In parallel, massive investment in sea transport, air traffic and related infrastructures, together with further liberalisation of foreign trade, will make the country the world's biggest trading nation.[53]

The flipside of China's strong industrial growth is its neglect of agriculture. Although most people in the mainly rural provinces are still involved in agriculture, domestic production in many regions is insufficient to feed the local population. On top of that, 200,000 hectares of farmland are being lost each year. Consequently, China will increasingly depend on agricultural imports.[54]

Moreover, enormous (and hardly affordable) investments in energy supply and power generation will be necessary to sustain the development of the industrial sectors. Further economic problems range from bad loans in the banking system – a problem that is being addressed – and bankrupt local governments to corruption and nepotism, which are closely related to the ongoing entrenchment of the state in the economy.[55] The discrepancy between underdevelopment and poverty in rural regions, on the one hand, and relatively high living standards in urban areas of the booming costal regions, on the other, is expected to grow further and may increase social tensions.[56]

Demographic trends are likely to reinforce these problems: population ageing will lead from 2015 onwards to a rapidly rising number of pensioners, expanding from 7% today to 20% (or 300 million) in 2025. Considerable resources will need to be directed to setting up a credible pension system for an elderly population lacking adequate pension insurance and social security.[57] Looming over all of this is the spectre of environmental degradation: by 2030, the direct impact of pollution on its population's health alone could cost China 15% of its GDP.[58] All this impedes China's development and creates a considerable potential for social unrest.

India

India is expected to be one of the fastest growing economies (maintaining its current growth pace of roughly 6% to 8% per year) and compete with Japan for the place of the third largest national economy in the world by 2025, behind the US and China. Favourable demographics, increasing (although as yet inadequate) investment in education and infrastructure and further integration into the world economy are the factors driving this process.[59] GDP per capita, however, will still be modest (at 12% of the US level in PPP in 2020). In fact, although this share has been steadily shrinking over the last few decades, about one third of India's population, or 360 million, still lives in conditions of extreme poverty with less than $1 per day. These figures put in perspective the remarkable economic achievements in specific sectors, and highlight the future challenges of development and employment. In this context, it should also be noted that India, thanks to its large diaspora in the US and the Gulf region, is among the biggest beneficiaries of remittances in the world, totalling $21.7 billion or 3.6% of GDP.[60]

Thanks to its young and rapidly growing population, India will add about 15 million workers to its labour pool each year over the next two decades. The country can draw on an enormous and ever-growing pool of (English-speaking) scientists, IT specialists, technicians and engineers which will further drive its economic development, although only a minority proportion of them is considered competitive by international standards.[61]

India has established itself as the world's leading location for IT services and business process outsourcing, with 25% of the world's offshored IT and IT-enabled services located there. Since many industrial countries have only just begun to shift some of their business services abroad, this sector is expected to grow further.[62] However, India will only be able to provide enough jobs for the younger generations and to reduce its still heavy dependence on agriculture (22% of GDP, employing 55% of the workforce) if it succeeds in developing a major industrial pillar. The textile and clothing industry, pharmaceuticals, civil aviation and biotechnology are considered to be among the most promising sectors. The main prerequisite for developing the industrial sector is to improve the country's largely inadequate physical infrastructure. In so far as transport infrastructure is concerned, estimated costs range between $150 and $200 billion over the next 10 to 15 years, although the costs involved to upgrade physical infrastructure at large could be much higher, up to $100 billion per year by 2010. Given the already high budget deficit, private investments, and particularly FDI, will be needed to finance big infrastructure projects.[63]

India is increasingly integrated into regional trade patterns. In 2004, ASEAN +3 (China, Japan, South Korea) nations have become, as a group, the biggest commercial partner of India, accounting for 20% of its trade. The EU follows closely at 19%, with the US remaining the single biggest national partner with 11% of total trade in 2004/2005 and China featuring second, following a stunning 80% increase in the volume of trade between the two countries over the same year.

Other parts of the world

Other Asian countries, in particular Malaysia, Thailand, Indonesia and Korea are also expected to post high GDP growth rates. The same is true for certain Latin American countries, such as Mexico, Chile, Argentina and Brazil.[64] Most other regions, in contrast, are expected to experience great difficulties in integrating into world markets and risk being left by the wayside.

Sub-Saharan Africa

Sub-Saharan Africa's share of world trade and foreign direct investment has fallen to a historical low of 2%. At the same time, the percentage of the population living with less than $1 per day has increased to a historical high of 46%. In most countries, economic growth rates fall short of the 7% GDP increase necessary to achieve the UN's Millennium Goal One of halving the number of poor people by 2015.[65] The continent, however, is not entirely stalled from an economic standpoint. In 2005, economic growth in Sub-Saharan Africa rose to 5.2% (against an average of 2.3% during the 1990s), which sends an encouraging signal for the future.

This positive performance needs, however, to be put in perspective. With a fast expanding population, GDP per capita has actually been decreasing in Sub-Saharan Africa by 1.3% over the 1980s and by 0.3% over the 1990s. Even if this trend were to be durably reversed, it will nevertheless take decades to overcome widespread poverty.

In addition, the economic performance of resource-rich and resource-poor countries largely differs: in 2005, oil-exporting countries, such as Nigeria, grew by 6.4%, and oil importing countries by only 4.3%. Many resource-rich countries remain heavily dependent on the export of a few primary commodities, including most notably oil but also iron, copper and diamonds, and have great difficulty diversifying into market-dynamic products. Rising oil prices, however, will be detrimental for poor importing countries, whose current account deficit has plunged.[66] Furthermore, inadequate infrastructure will impose serious constraints on the energy supply of these countries, and undermine their potential for growth.

African agriculture is undermined by lack of irrigation, lack of investment and worsening environmental conditions. Productivity is low and, as a consequence, food imports are expected to significantly increase. Turning to manufacturing, progress can be detected in specific sectors, such as textiles, but a lack of economies of scale and growing competition from emerging countries will likely narrow the margins for growth. On top of that come low savings and investment rates, as well as the low quality of infrastructure and underdeveloped human capital. Far-reaching economic and governance reforms are necessary to improve this situation. However, the capacity to implement them will likely differ greatly between African countries.

Thanks to the existence of an indigenous business class and a well-developed infrastructure, South Africa is expected to perform much better than any other Sub-Saharan country. Economic spillover could also promote growth in neighbouring countries and foster development in Southern Africa in general. On the other hand, South Africa's economy is handicapped by the small size of its domestic market and its geographical distance from world markets. High AIDS rates also hamper further economic development.[67]

On the whole, leaving aside the pressing challenge of poverty reduction, the economic development of most African countries will mainly depend on the capacity to attract foreign investment, retain skilled manpower, build infrastructure and take advantage of technological developments. The weaker the governance record, the greater the risk for these countries of being marginalised and falling victim to the dark side of globalisation.[68]

The Middle East and North Africa

MENA is a very heterogeneous region, with important differences in particular between resource-poor countries (Egypt, Jordan, Lebanon, Morocco and Tunisia), labour-abundant, resource-rich countries (Algeria, Iran, Iraq,

Yemen and Syria) and the labour-importing, resource-rich GCC states.[69] On the whole, all MENA economies have underperformed in the past decades, with GDP, productivity and investment rates well below the global average. It is therefore generally recognised that the dominant economic model of the region – based on the public sector, oil incomes and workers' remittances — is not up to the challenges of globalisation.

According to the World Bank, the region needs in particular to shift its sources of growth – from oil to non-oil sectors, from state-dominated to private, market-oriented investment, and from protected import substitution to export-oriented activities.[70] That said, an important distinction should be drawn between resource-rich and resource-poor countries. The former have benefited from booming oil prices, making up for about 40% of their GDP since 2002. In this group, GCC countries have made progress in mobilising the oil wealth to modernise their economies and their infrastructures, whereas so far countries such as Iran and Algeria have proved less successful at fostering significant economic diversification.

Economic reforms are urgently needed, in particular to provide jobs. Population growth in the region implies an enormous increase of the workforce to some 185 million people by 2020. Under a worst case scenario, this could bring unemployment up from 15 to 50 million over this period, affecting in particular young and female jobseekers with intermediate and higher education qualifications. To absorb all this labour, MENA economies would have to maintain investment rates of 30% of GDP and annual income growth of 6-7% – which is far beyond their recent record.[71] Even the resource-rich GCC states may have great difficulties in generating adequate employment opportunities for young job seekers (who constitute nearly 25% of the population).[72]

Expanding trade and attracting foreign capital will be increasingly difficult, since the region faces growing competition in world markets both at the skill-intensive end (from EU candidate countries and NAFTA among others) and at the labour-intensive end (from low-wage, high productivity countries, such as Bangladesh or Indonesia).[73] FDI, which amounted in 2005 to $9.1billion and accounts for 0.9% of global FDI (3.8% of FDI to developing countries),[74] is mostly directed to the energy sector.

Whether MENA countries will be able to tackle these challenges will depend both on internal political reforms and developments in the region as a whole. In many countries, strong resistance persists against transition from state-dominant and protectionist to open, market-led economic systems. Positive developments can be seen in Morocco and Tunisia, as well as in the smaller GCC states (in particular the UAE). However, economic prospects in the region will remain closely linked to political developments (i.e. in the context of the Israeli-Palestinian conflict, Iraq and Iran).

Russia

Between 2003 and 2005, Russia posted GDP growth rates ranging between 7.3% and 6.4%, with GDP rising to $581 billion. Important economic and fiscal reforms have been recently introduced, which improve the governance framework and pave the way for sustainable growth. According to some projections, Russian GDP would maintain high growth rates over the next 20 years, and approach the level of France's or Italy's.[75] Economic growth is likely to be driven mainly by the increase in energy prices, which should benefit Russia as a major exporter of oil and gas.[76]

Three factors, however, put the sustainability of high growth rates into doubt. First, Russia's growth largely depends on energy exports, while economic diversification has not been seriously undertaken, and tight public control over the energy sector will likely entail poor management and collusion. Second, it is estimated that huge investments will be needed to maintain and expand energy infrastructure, quantified at more than €700 billion between 2003 and 2020.[77]

Third, the investment climate is likely to remain unfavourable for the foreseeable future. The poor quality of its legal system and public administration, the selective application and enforcement of laws and the high level of crime and corruption are expected to remain major impediments to foreign investments and economic growth. FDI directed to Russia stood at $9.4 billion in 2004, and the accumulated stock of FDI in the country as a share of GDP amounts to only a fifth of the level of other transition economies in Europe.[78] At the same time, the country has one of the highest rate of HIV infections in the world, which affects the younger generation in particular and thus impacts on labour supply and productivity.[79]

Notes

1. *Foresight 2020: Economic, Industry and Corporate Trends*, Economist Intelligence Unit, March 2006.
2. In the case of the pharmaceutical industry, for example, India's actual specialisation in generic drugs could progressively shift to primary research and medical technology. 'Dynamic sectors give global growth centres the edge', Deutsche Bank Research, Frankfurt, October 2005.
3. 'Trade and Structural Adjustment: Embracing Globalisation', OECD, Paris, 2005.
4. Relocation of manufacturing industries to developing countries will mostly affect industrial sectors where production is labour-intensive, leads to economies of scale, and faces low transport costs. See *Foresight 2020: Economic, Industry and Corporate Trends*, op. cit.
5. *World Investment Report: The Shift Towards Services*, UNCTAD, United Nations, 2004.
6. A firm can offshore an activity either by producing through a foreign affiliate (FDI) or by externalising to another firm abroad. See *World Investment Report: The Shift Towards Services*, op. cit.
7. 'FDI in developing countries takes off: is a new geography of investment emerging?', UNCTAD, Press release, 10 August 2004.
8. One of the key factors driving the so-called South-South FDI is the strategic goal to secure the provision of raw materials, including oil and gas. See Dilek Aykut & Dilip Ratha, 'South-South FDI flows: how big are they?', *Transnational Corporations*, vol. 13, no. 1, April 2004.
9. 'Asian Development Outlook: routes for Asia's trade', Asian Development Bank, 2006.
10. As a result, intra-regional Asian trade is increasingly composed of parts and components (more than 60% in 2003 from less than 50% in 1993).
11. Dominic Wilson, Roopa Purushothaman and Themistoklis Fiotakis, *The BRICs and global markets: crude, cars and capital*, Goldman Sachs Global Economics Paper no. 118, October 2004.
12. Existing production capacities will nevertheless remain located in the EU, the US and Japan, because final assembly is a capital-intensive process. However, the production of components could shift towards these markets. See *Foresight 2020... op. cit.*
13. Currently, 197 preferential arrangements notified to the WTO are in force, and around 70 are under negotiation/at proposal stage. At the creation of the WTO in 1995, only 50 RTAs notified to the WTO were in force. And between January 2004 and February 2005 alone, an unprecedented 43 new RTAs have been notified to the WTO. See Jo-Ann Crawford & Roberto V. Fiorentino, 'Changing the landscape of regional trading agreements', World Trade Organization Discussion Paper no. 8, 2005.
14. 'Extreme poverty, defined by the World Bank as getting by on an income of less than $1 a day, means that households cannot meet basic needs for survival. (...) Unlike moderate or relative poverty, extreme poverty now exists only in developing countries. Moderate poverty, defined as living on $1 to $2 a day, refers to conditions in which basic needs are met, but just barely.' Jeffrey D. Sachs, 'The end of poverty', *Time Magazine*, 6 March 2005.
15. In 2002, however, 437 million people still faced extreme poverty in South Asia, which amounts to 31% of the population of the region. South Asia remains the region with the largest pool of people living in extreme poverty in the world, in absolute terms. See *World Development Report 2006: Equity and Development* (The World Bank, 2005).
16. On the other hand, the number of people living on less than a dollar per day is expected

to decrease to only 11 million in China and 232 million in South Asia (and from 42 to 29 million in Latin America). 'Global Economic Prospects: Economic Implications of Remittances and Migration', The World Bank, 2006.

17. The number of people living on less than $2 per day increased not only in Africa, but also slightly in South Asia (from 996 to 1091 million). In 2002, it represented 75% of the Sub-Saharan population and 78% of the population in South Asia.

18. Where the number of people living on less than $2 per day is expected to drop from 533 to 181 million in less than ten years from now.

19. Even if not particularly damaging for the whole economy, shifts in the international location of international production can exhibit considerable adjustment costs for specific industries or geographical regions, raise serious concerns in affected communities and set the political debate off on a defensive and protectionist track.

20. Developing countries generally maintain high tariff levels and revenues from import duties are very considerable, especially for the least developed countries who lack an efficient tax regime and a large domestic tax base. Import duties represent on average 18% of total government revenues in low income countries, and as much as 34% in the least developed countries in Africa over the period 1999-2001. Unless these countries succeed in replacing import duties by other revenues such as taxes on income, sales or value added, the ability of governments to mobilise resources and support the necessary diversification of their economy will be compromised. See 'Trade and Structural Adjustment: Embracing Globalisation', OECD, Paris, 2005.

21. It should be noted that, at current prices, the share of world GDP produced by the US, the EU and Japan stood as high as 70% in 2005.

22. See Dominic Wilson and Roopa Purushothaman, *Dreaming with BRICs: The Path to 2050*, Goldman Sachs, Global Economics Paper no. 99, 1 October 2003.

23. The present GDP ranking, at current prices, is the following, from the top down: the US, Japan, Germany, the UK, France, China and Italy.

24. GDP per capita is calculated here in PPP because it gives a better reflection of the real standard of living of the people in any given country.

25. *Foresight 2020...*, op. cit.; and *Dreaming with BRICs: The Path to 2050*, op. cit.

26. *The Transatlantic Economy in 2020: A Partnership for the Future?*, The Atlantic Council, Policy Paper, Washington, November 2004.

27. Giovanni Dosi, Patrick Llerena, Mauro Sylvos Labini, *Evaluating and comparing the innovation performance of the United States and the European Union*, Expert Report for the Trend Chart Policy Workshop 2005.

28. John Williamson, 'What Follows the USA as the World's Growth Engine?', India Policy Forum Public Lecture, July 2005.

29. Alan J. Auerbach, William G. Gale, and Peter R. Orszag, 'Sources of the long term fiscal gap', Tax Note, 24 May 2004, Tax Policy Center.

30. See *World Economic Situation and Prospects 2006* (United Nations, New York, 2006); *The Long-Term Budget Outlook* (Congressional Budget Office, December 2005); Barry Eichengreen: 'The Blind Men and the Elephant', *Issues in Economic Policy*, no. 1, January 2006 (The Brookings Institution, Washington).

31. *The Transatlantic Economy in 2020: A Partnership for the Future?*, op. cit.

32. Annual negative impact on GDP of the loss of working force is estimated at an average of 0.5% from 2010 to 2030, even if it seems likely that growth of productivity and/or working time could mitigate this trend. See Klaus Regling, 'How Ageing Will Torpedo Europe's Growth Potential', *Europe's World,* Spring 2006; *Global Economic Prospects: Economic Impli-*

cations of Remittances and Migration, op. cit.; *The Impact of Ageing on Public Expenditure: Projections for the EU 25 Member States on Pensions, Health care, Education and Unemployment Transfers (2004-2050)*, Special Report no. 1/2006, Economic Policy Committee and the European Commission (DG ECFIN).

33. To date, there is little precise estimate of the effect of delocalisation. Germany is estimated to have lost 90,000 jobs over 1990-2001 due to the competition of new EU Member States of Central and Eastern Europe, only 0.3% of total jobs and a small fraction of the number of jobs being lost each year in Germany. See Lionel Fontagné and Jean-Hervé Lorenzi, 'Désindustrialisation, délocalisations', *Rapport du Conseil d'Analyse Economique*, no. 55, October 2004.
34. 'Dynamic sectors give global growth centres the edge', op. cit.
35. CEPII-CIREM, 'European industry's place in the international division of labour: situation and prospects', Report prepared for the Directorate-General for Trade of the European Commission, July 2004.
36. Daniele Archibugi and Alberto Coco, 'Is Europe becoming the most dynamic knowledge economy in the world?', *Journal of Common Market Studies*, vol. 43, no. 3, September 2005.
37. Wide differences in R&D expenditures and scientific output exist among EU Member States, with for instance Sweden (3.98%) and Finland (3.49%) leading OECD countries in terms of R&D intensity. See *OECD Science, Technology and Industry Scoreboard 2005* (OECD, Paris, 2005).
38. Only 55.6% of global R&D in the EU is financed by the private sector, as compared with 63.1% in the US and 73.9% in Japan, which is again far from the target agreed at Lisbon in 2000 (two thirds).
39. European Commission, 'Towards a European Research Area: Science, Technology and Innovation – Key Indicators 2005', 2005.
40. *Long-term Economic Forecast of the Japanese Economy* (2001-2025), Japan Centre for Economic Research, Tokyo 2001.
41. *Foresight 2020 ...*, op. cit., and Goldman Sachs, *Dreaming with BRICs: the Path to 2050*, op. cit.
42. 'Economic Survey of Japan, 2006', Policy Brief, OECD, July 2006.
43. Ibid.
44. In particular on the revenue side, a broadening of the tax base or an increase in the consumption tax rate could be necessary. In addition, social security and medical care expenditures are currently, even after the recent pension reform, projected to reach 20% of the GDP in 2025, against 16% today. See Daniel Citrin and Alexander Wolfson, 'Japan's Back', *Finance and Development*, vol. 43, no. 2, June 2006.
45. 60% of Japan's exports to China are for processing. See Guillaume Gaulier, Françoise Lemoine & Deniz Ünal-Kesenci, 'China's emergence and the reorganisation of trade flows in Asia', CEPII Working Paper, May 2006.
46. Legal and illegal foreign workers represent a meagre 1% of the employed workforce. See 'Economic Survey of Japan, 2006', op. cit.
47. 'Growth and Trade Horizons for Asia: Long-term Forecasts for Regional Integration', ERD Working Paper Series no. 74, Asian Development Bank, Manila, November 2005.
48. Stefan Bergheim, *Global Growth Centres 2020*, Current Issues Working Paper, Deutsche Bank Research, Frankfurt, March 2005.
49. Carsten A. Holz, *China's Economic Growth 1978-2025: What We Know Today about China's Economic Growth Tomorrow*, Centre on China's Transnational Relations, Working Paper no. 8, The Hong Kong University of Science and Technology, July 2005.

50. In the early 1980s, raw materials and oil accounted for a 50% share of Chinese exports; in the early 1990s, garment production, with massive investments in textile plants, became the main engine of growth. Since the mid-1990s, the principal export chapters were auto parts and consumer electronics. The chemical and automobile industry are expected to be future growth sectors. See: 'Dynamic sectors give global growth centres the edge', op. cit.
51. See *UNESCO Science Report 2005* (UNESCO, Paris 2005).
52. Carsten A. Holz, op. cit.
53. See 'Dynamic sectors give global growth centres the edge', op. cit.
54. Ibid.
55. See Minxin Pei, 'The Dark Side of China's Rise', in *Foreign Policy*, March/April 2006.
56. Andreas Hoffbauer, '800 Millionen haben nichts von Chinas Boom', *Handelsblatt*, 2 February 2006.
57. Finn Mayer-Kuckuck, 'Der Drache wird alt, bevor er reich wird', *Handelsblatt*, 7 February 2006; Petra Kolonko, 'Wohin mit Chinas Alten', *Frankfurter Allgemeine Sonntagszeitung*, 5 March 2006.
58. Jonathan Shaw, 'The Great Global Experiment', *Harvard Magazine*, vol. 105, no. 2, November/December 2002, p. 87.
59. 'India rising: A medium-term perspective', Deutsche Bank Research, Frankfurt, May 2005.
60. 'Global Economic Prospects: Economic Implications of Remittances and Migration', op. cit.
61. Ibid.
62. 'Human capital is the key to growth', Deutsche Bank Research, Frankfurt, August 2005.
63. *India Vision 2020*, Planning Commission, Government of India, New Dehli 2002; Chetan Ahya and Mihir Sheth, 'India: Infrastructure: Changing Gears', Morgan Stanley Report, 2005.
64. 'Global Growth Centres 2020', op. cit.
65. UN Economic Report on Africa, November 2005.
66. In 2005, oil-exporting countries experienced a rising trade surplus (19.8% of GDP) whereas non-oil exporting countries experienced a deepening of their deficit (6.6% of GDP). *African Economic Outlook 2005/2006* (OECD, Paris, 2006).
67. Between 1992 and 2000 the HIV/AIDS epidemic reduced the growth of GDP per capita for 33 African countries by 0.7% per year. Were the epidemic to maintain the same intensity between 2002 and 2020, it is estimated that Sub-Saharan Africa would grow 18% less – a loss of $144 billion. The impact of HIV/AIDS on the personnel of business and public structures, as well as on the costs of health services and on measures for poverty alleviation, should be added to these figures. See the *Economic Report on Africa: Meeting the Challenges of Unemployment and Poverty in Africa* (Economic Commission for Africa, United Nations, 2005).
68. 'Mapping Sub-Saharan Africa's Future', National Intelligence Council, Discussion Paper, March 2005; 'Enhancing the Contribution to Development of Indigenous Private Sector in Africa: Challenges and opportunities for Asia-Africa cooperation', Afrasia Business Council, 2005.
69. 'Middle East and North Africa. Oil booms and revenue management: Economic Developments and Prospects 2005', Report on the Middle East and North Africa Region (The World Bank, Washington D.C., 2005).
70. 'Trade, Investment, and Development in the Middle East and North Africa – Engaging

71. 'Unlocking the employment potential in the Middle East and North Africa – Towards a new social contract', *MENA Development Report* (The World Bank, Washington D.C., 2005).
72. Marcus Noland and Howard Pack, 'Islam, Globalisation and Economic Performance in the Middle East', in: *International Economics Policy Briefs*, June 2004.
73. 'Trade, Investment, and Development in the Middle East and North Africa – Engaging with the World', op. cit.
74. *Global Development Finance: The Development Potential of Surging Capital Flows* (The World Bank, Washington D.C., 2006).
75. *Dreaming with BRICs: The Path to 2050*, op. cit.
76. 'Russia 2010: scenarios for economic development', Deutsche Bank Research, Frankfurt, March 2003.
77. Christian Cleuntinx, 'The EU-Russia Energy Dialogue' (DG for Energy and Transport, European Commission: Vienna, December 2003).
78. EBRD estimates, quoted in the report *Russia: Investment Destination* (Foreign Investment Advisory Council, March 2005): http://www.pbnco.com/fiacsurvey/.
79. See 'The Economic Consequences of HIV in Russia', The World Bank, Russia Office, 10 November 2002.

Energy

3

Energy demand, supply, consumption and efficiency will be key variables affecting economic growth and political stability. With demand growing faster than supply, and the price of oil and gas rising for the foreseeable future, access to energy resources will be subject to growing geopolitical competition. Massive investment will be required to maintain and expand existing infrastructures. Technological innovation will be essential in harnessing the potential of renewables, at a time when the finite nature of fossil fuels reserves is becoming tangible, and our model of economic development is attaining the limits of environmental sustainability.

General Trends

- Between 2006 and 2025, global demand for primary energy is expected to grow at an average rate of 1.6% per year.[1] By around 2030, energy requirements are predicted to be more than 50% higher than today.[2]

- Fossil fuels (oil, gas and coal) will continue to be the world's primary energy source, accounting for 81% of demand.[3] Oil will remain the single largest energy source while natural gas will overtake coal and rank second by around 2020.[4]

- The share of nuclear energy is expected to decline in most mature market economies, but to increase in developing countries.[5] Renewable energies will grow faster than any other energy source, in particular in OECD countries, but will still make only a small contribution to total global energy supply.[6]

- Developing countries will account for more than two thirds of the increase in energy demand. Regional disparities will nevertheless be important, the growth (in volume terms) in Asia being much higher than in Africa.[7] Demand will also grow in the OECD world, but at a

> lower rate.[8] The dependency of developed and emerging countries on energy imports will increase substantially.
>
> ▶ Energy resources are most probably sufficient to meet the projected growth in demand, but their effective exploitation is strictly conditioned to investments. In addition, while both energy demand and energy supply will increase, the former will grow faster than the latter, which could result in higher energy prices.

According to the IEA reference scenario 2005, global energy demand is projected to increase by 52% from now to 2030, reaching 16.3 billion tonnes of oil equivalent (btoe). Fossil fuels will make up the biggest share of the increase in the demand (81%).[9] Due to growing energy demand, potential political instability in many supplier countries, but also possible investment shortfalls, oil and gas prices are expected to increase over time.[10]

Oil will remain the single largest source of energy, and demand is expanding fast.[11] Between 2001 and 2005, world oil demand increased by 8.8% per year, but China's consumption alone went up 46%.[12] Between 2004 and 2030, world oil consumption is expected to increase by 40%, from 82.1 million barrels a day (mb/d) to 115.4 mb/d.[13] Two thirds of this increase will go to meet transport sector demands.[14] The share of oil in global energy demand is expected to remain roughly the same (35%).[15] According to most estimates, the size of existing reserves and the discovery of new fields will be sufficient to meet demand for the two decades to come. Leaving aside political tensions and the consequent potential disruption of flows, however, some question whether supply will keep up with the accelerating pace of demand. The discovery of new fields has been declining for years, and a large share of the world's production relies on a few giant fields, notably in the Middle East.[16] The capacity of Caspian reserves remains uncertain but expectations have been reappraised. The OPEC considers that the production of non-OPEC countries will reach a plateau around 2015 at 58-59 mb/d, with the bulk of the increase coming from offshore fields in Africa and Latin America.[17]

Global consumption of **natural gas** will increase by 87%, from 2,622 to 4,900 billion cubic metres (cm) in 2030.[18] The market for liquefied natural gas (LNG) is expected to grow even more rapidly, with demand expanding by about 10% a year.[19] The share of gas in world energy demand will rise from 20% in 2003 to 24% in 2030.[20] Tensions on the gas market can be envisaged, stemming from the gap between growing demand and inadequate infrastructures to deliver supply, be it by land or by sea.

Coal consumption is expected to fall in mature economies, but to increase in developing countries. Demand is projected to grow worldwide from 5,200 million tonnes (Mt) in 2003 to almost 7,300 Mt in 2030.[21] At a time when oil and gas supplies are subject to growing uncertainty, and their price is mounting, coal can provide a relatively cheap and attractive alternative for both developed and developing countries. While coal remains a relatively polluting energy source, new technologies make coal exploitation more environmentally sustainable, which is part of the reason why the orders for coal-based power plants have recently outpaced those for gas-based plants.[22] Today, two thirds of coal is used for power generation, and coal represents 40% of worldwide electricity production.[23] Power generation will remain the main driver of coal demand, notably with a view to the growth of electricity consumption in emerging countries such as China and India.[24] These countries hold abundant reserves of coal and are distant from gas suppliers. On the whole, unless access to other energy sources becomes too expensive, the coal share of world energy demand is expected to fall slightly (from 24% to 23%).

The share of **nuclear power** in global energy demand is expected to decline from 6.4% in 2003 to 4.7% in 2030,[25] with production peaking in 2015. The largest decline will occur in Europe (provided attitudes and politics towards nuclear energy do not change), whereas nuclear output will increase in Asia.[26] Today, nuclear energy accounts for 17% of global electricity consumption. Electricity demand may grow by 100% between now and 2030,[27] which should entail renewed interest in the potential of nuclear energy for non-polluting power production. China and India, in particular, have envisaged considerable investments in new nuclear power plants, although those will cover only a modest proportion of their sky-rocketing electricity needs. In this context, the European Commission has recently sought to relaunch the debate on investment in the nuclear sector, but there is a clear understanding that decisions pertain to individual EU Member States, whose policies widely differ.[28] However, if power production is to be environmentally sustainable over the long term, notably with a view to reducing greenhouse gas emissions levels, and if dependency on energy supply is to be curbed, the debate on nuclear energy will need to be tackled again at some point in the future.[29] The recent policy paper unveiled by the British government, envisaging that investment on renewables and energy efficiency could be flanked by the re-launch of its nuclear programme, is a telling illustration of this.[30]

The share of **biomass** in world energy demand will remain at roughly 10% in 2030. In developing countries, traditional biomass will increasingly be replaced by modern commercial fuels, such as ethanol, but its consumption in absolute terms will continue to grow. New biofuels can potentially provide a viable alternative to oil for the transport sector, as the Brazilian but also the German experiences show, notably when oil prices are high. The price of fossil fuels, technological innovation and the required investment in infrastructure will be among the key factors determining the shift towards biofuels. In developed countries, commercial biomass and waste will be increasingly used in power generation. According to the European Environmental Agency, the bio-energy potential in 2030 will represent around 15-16% of the projected primary energy requirements of the EU-25.[31]

Hydropower represents the main source of renewable power generation. The EU as well as the US capacity of hydropower production has nearly reached its peak. On the other hand, hydropower will be the main driver of Chinese power generation increase, far ahead of the nuclear sector. Unexploited potential still exists in Latin America, India and, to a lesser extent, Africa. The use of solar energy and wind power will grow rapidly (+365% between 2002 and 2030), in particular in developed countries, but also in

Primary energy demand in main developed and emerging economies (in million tonnes of oil equivalent)

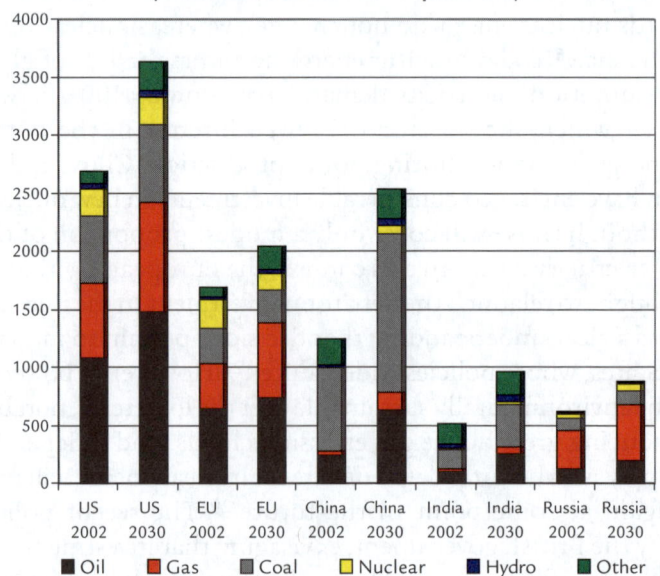

Source: Energy Information Administration, *International Energy Outlook 2006*, US Department of Energy, June 2006.

emerging and poor countries, with considerable potential for hydropower in Africa and India. However, their share in primary world energy consumption is expected to be still modest, at 2% in 2030. On the whole, the share of sustainable renewables, except biomass (hydropower, biofuels, solar energy, wind power etc.) is expected to stand at 8% of global energy consumption. Traditional biomass, such as wood burning, will, remarkably, still account for 6% of total energy consumption.[32]

Driven by population and economic growth, urbanisation and industrialisation, **developing countries** will make up for more than two thirds of the increase in world energy demand.[33] Mainly because of the booming demand in Asia, in particular China and India,[34] the developing world's oil and coal consumption is projected to almost double, and its gas consumption to almost triple.[35] Increase in energy demand will be less pronounced in the OECD world. However, in 2030, developed countries are still expected to consume more oil (55.1 vs. 50.9 million b/d) and gas (2,061 vs. 1803 billion cm) than developing countries.[36] The same is true for nuclear energy as well (2,083 vs. 675 billion kWh in 2025), in spite of the envisaged decline in consumption in most developed countries. Only the consumption of coal will be higher in developing countries (3,984 Mt vs. 2461 Mt in 2030).[37]

Increase in primary energy production will occur almost exclusively in the non-OECD world (the Middle East, Russia, Latin America, Africa). In the US, the domestic production of oil is expected to decline (from 5.4 mb/d in 2004 to 4.5 mb/d in 2030) [38] but gas production should increase, whereas in Europe offshore oil resources are quickly depleting. Other developed and emerging countries do not dispose of significant resources (with the exception of coal). As a consequence of these trends, and of the booming demand for energy, exports of oil and gas from developing to developed and emerging countries will increase significantly, leading to a situation of acute **dependency on energy imports**. By the horizon year 2025 Europe is expected to depend on external supply for 90% of its oil needs and 80% of gas, and these shares will increase further to 94% and 84% in 2030.[39] Dependency will be less absolute, but nevertheless large and growing, for the US, which will import 51% of its oil in 2020 and up to 66% in 2030 (as opposed to 47% in 2004).[40] Between 2020 and 2030, India's dependency on oil imports will increase, from 87% to 91% (69% in 2002). As far as China is concerned, in the same period, the oil dependency will rise from 75% to 84% (34% in 2002).[41] By 2030, India and China will need to import, respectively 40% and 27% of their gas.[42]

The New Global Puzzle — PART I

Source of oil supply for the US, EU and China in million barrels per day (mb/d)

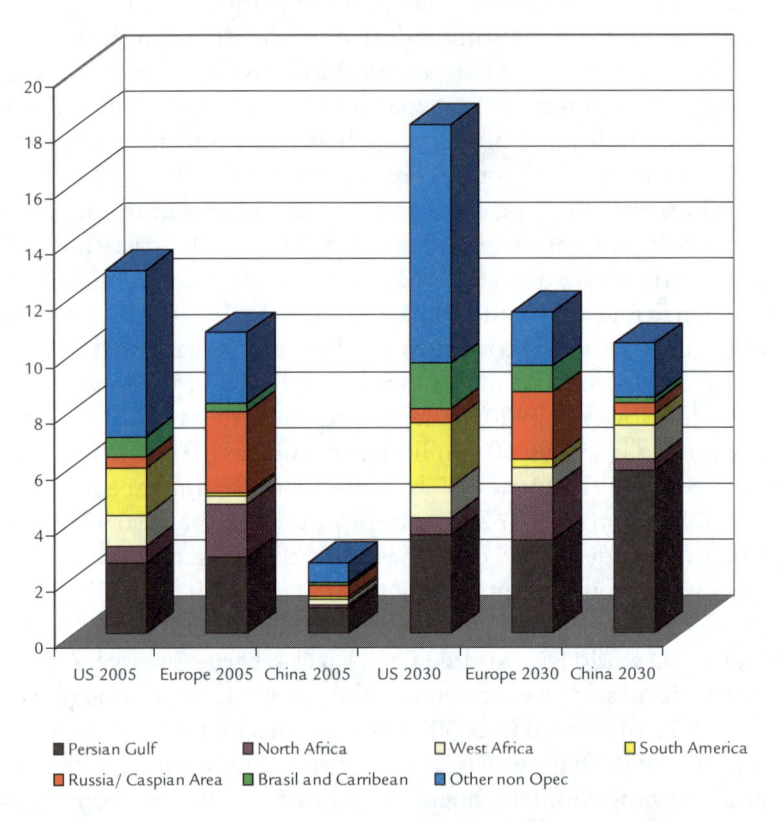

[Bar chart showing oil supply sources for US 2005, Europe 2005, China 2005, US 2030, Europe 2030, China 2030, broken down by: Persian Gulf, North Africa, West Africa, South America, Russia/Caspian Area, Brasil and Carribean, Other non Opec]

Source: Energy Information Administration, *International Energy Outlook 2006*, US Department of Energy, June 2006

As a result of these imbalances, importing countries will seek to diversify their suppliers, and exporting countries will be in a better position to select their clients. The success of both strategies will notably depend on existing and envisaged infrastructures, among other factors. Under all circumstances, importing countries will need to enter tougher competition to ensure their supplies. At the same time, however, dependency works both ways, with many producers relying heavily on energy exports for their economic development and political stability. In addition, when it comes to the oil market, importing countries are exposed not only to the sudden disruption or reduction of supply flows directed to them but also, more likely, to turbulences in supply anywhere in the world. In a global market, these would impact on the price of oil regardless of the geographical location of the crisis spot.

An estimated total of $16.4 trillion will have to be invested in infrastructures until 2030 to meet the growing energy demand ($6 trillion in oil and gas, $400 billion in coal, and $10 trillion in power generation). Whether the necessary **investments** will be made in time to avoid a bottleneck in supply remains to be seen. The replacement of existing infrastructure will consume the bulk of envisaged investment for the energy sector – around 51%.[43] Considerable investment will also be required in maritime transport – due to grow by over 93% in the next 25 years in terms of oil tankers' capacity alone[44] – and oil refining, notably to meet the demand of large emerging markets. In the gas sector, investment in infrastructures should nearly add one million km to the existing pipeline network of 1.1 million km,[45] and maritime transport capacity is expected to double. Compared to these ambitious targets, perceptible under-investment in physical infrastructures may lead to future shortages.[46]

Some experts, moreover, fear a hiatus between demand and supply between 2015 and 2020, due to little progress in the discovery and exploitation of new reserves. Massive investment would be required in the Eastern and Arctic regions of Russian Siberia but also in Canada, Venezuela and Africa. Given pressing needs and a mounting energy demand, investment will concentrate on the most directly profitable projects. Those developing countries deprived of resources risk failing to attract the necessary foreign investment to renew their own infrastructures. Of the world's 47 poorest countries, 38 are net oil importers, and 25 import all of their oil needs. Together with potential political instability and/or national market rigidities, inadequate investment may well lead to serious supply problems. According to the UN, limitations on energy access may become one of the major obstacles to economic growth in developing countries.[47]

Resource nationalism may pose yet another obstacle to investment.[48] According to the IEA, around 57% of world oil reserves are shielded from foreign access, with priority given to national companies. The latter enjoy an exclusive position in Saudi Arabia, Kuwait and Mexico, home to 35% of the proven reserves of oil. Moreover, former liberalised markets, such as the Latin America oil and gas market, are increasingly locked by national authorities and governmental interference is on the rise in Russia, Venezuela, Bolivia and Ecuador. In South America more precisely, energy is perceived as a means of development (Bolivia, Ecuador and Argentina) but also as a vector to counterbalance the overwhelming presence of the US and of certain foreign corporations (Venezuela, Bolivia). These recent trends imply that a substantial share of investment will be diverted from the energy sector, entailing in turn a slowdown of expected production growth.

In parallel, the introduction of new actors, such as Chinese companies, exacerbates the competition between energy companies even more and increases the access costs to new oil and gas fields accordingly.

Regional differences

Major energy exporters

OPEC countries

The Middle East will continue to play a pivotal role in world energy supply. Oil reserves in the region are the biggest in the world (expected to last for over 70 years), of high quality and easily exploitable.[49] The Middle East currently produces 28% of the world's oil. This share should rise to 43% by 2030. By 2030, oil production from the entire MENA region is expected to increase by 74% from 29 mb/d to 50.5 mb/d, and gas production to triple and reach 900 bcm in 2030.[50] At the same time, the Middle East has enormous proven natural gas reserves (expected to last 245 years), and its gas production is expected to triple by 2030.[51] Around 50% of both China's and India's oil demand is supplied by the Middle East, while the flow of gas from Iran and other Gulf countries to China and India might grow via the envisaged (but highly controversial) overland pipeline across Pakistan and/or in liquefied form, through the Indian Ocean.

Saudi Arabia has the world's largest oil reserves and will remain the biggest oil exporter with an expected production increase of 75% by 2030.[52] With a notable shift compared to the last decades, the main export destinations will be Asia, followed by the USA and Europe.[53] At the same time, Saudi Arabia has the fourth largest proven reserves of natural gas in the world (6.7 trillion cm) after Russia (48 trillion cm), Iran (28 trillion cm) and Qatar (26 trillion cm).[54] Its gas production is expected to increase from 60 billion cm in 2003 to 155 billion cm in 2030.[55]

Iraq possesses one of the largest oil reserves in the Middle East.[56] The IEA forecasts Iraq 'to see the fastest rate of production growth, and the biggest increase in volume terms after Saudi Arabia'[57] (+119 % production increase by 2030).[58] Natural gas production is expected to grow sharply from 2 billion cm in 2003 to 32 billion cm in 2030.[59] However, these projections do not take into account the security context and the uncertain political future of the country.

Iran also holds enormous oil reserves and is expected to increase its production by 66% by 2030.[60] Currently, half of Iranian oil is exported to Europe, while the rest goes to Asia (South Korea, Japan and China).[61]

The world's biggest oil producers* and
estimates of proven crude oil reserves

Countries	Production 2004 (millions of b/d)	Production 2030 (millions of b/d)	Reserves (in 2003) billions of barrels**
Saudi Arabia	10.4	18.2	262
USA and Canada	9.7	7.4	27.2
Russia	9.2	11.1	60-69
Asia Pacific	6.2	4.3	39
OECD Europe	6.0	2.3	18 (Western Europe)
Iran	4.1	6.8	133
Latin America	3.8	6.1	117
Mexico	3.8	3.4	-
Non-OPEC Africa***	3.3	8	114
Venezuela	3.1**	5.6***	77
UAE	2.7	5.1	97
Kuwait	2.5	4.9	99
Iraq	2.0	7.9	115

* Source: International Energy Agency, *World Energy Outlook 2005*;
** Source: OPEC, *Annual Statistical Bulletin 2004*;
*** Source: Energy Information Administration, *International Energy Outlook 2006*.

Natural gas reserves (28 trillion cm) are the second largest in the world after Russia,[62] and gas production is expected to grow steadily from 78 billion cm in 2003 to 240 billion cm in 2030, mainly due to the exploitation of the giant South Pars field resources.[63] Iranian gas will be exported increasingly to Europe and to Asia, with new pipelines being planned or at least discussed.[64]

The smaller Gulf states also have important energy reserves. **Kuwait** and the **United Arab Emirates (UAE)** are projected to almost double their oil production to 4.9 and 5.1 million b/d respectively, exporting mainly to the Asian-Pacific region and Japan.[65] The UAE are also a major gas producer (44 billion cm in 2004) and hold proven gas reserves of 6.1 trillion cm (which are roughly equal to those of Saudi Arabia). **Qatar** has the third largest gas reserves in the world (26 trillion cm) and is expected to dramatically increase its gas production from 41 billion cm in 2004 to 255 billion cm in 2030 (representing 30% of total Middle East production), with a growing share in LNG.

Venezuela will remain the most important OPEC producer outside the Middle East, followed by Nigeria, Libya and Algeria.[66] By 2025, Venezuela's oil production is expected to increase by more than 80%, although ongoing political developments, which strengthen government control over energy resources, might dissuade foreign investment and therefore, in time, lead to a stagnation of production.[67] **Nigeria**'s production could double and reach 4.4 million by 2020. The US shows considerable interest in Nigerian oil, mostly because of its quality and of the absence of maritime chokepoints between the two countries.[68] **Libya** seems on the way to becoming another major supplier, with production expected to grow up to 3 mb/d in 2015.[69] **Algeria**'s oil production is projected to decrease somewhere around 2020. Its gas production, in contrast, which is the largest of all OPEC countries (88 billion cm in 2003), will double, with Europe as its main export destination.[70] The use of LNG and the construction of a third pipeline to Europe should boost the volume of exports (144 billion cm in 2030). Algerian gas reserves are estimated at 4.6 trillion cm.

Non-OPEC countries

Non-OPEC energy supply is expected to continue to increase as well,[71] with Russia and the Caspian region playing a special role. **Russia** holds roughly 6% of the world's oil reserves, mainly in Western Siberia (proven reserves are projected to last 19 years, recoverable reserves 2-3 times longer).[72] Its oil production is projected to increase by 22 % from 8.5 million b/d in 2003 to 10.4 million b/d in 2030.[73] However, lacking infrastructures and investment, it seems doubtful whether such growth rates will be sustainable in the long run.[74] In fact, according to some governmental sources, the peak in Russian oil production may be attained much earlier.[75] By 2025, Russian oil (which is of poorer quality compared to Arab oil) is expected to be exported mainly to Northeast Asia, with possible frictions between China and Japan as a result.[76]

At the same time, Russia is the world's largest gas exporter, with estimated reserves of 48 trillion cm.[77] Russian gas production (608 billion cm in 2003) is expected to expand further,[78] with export destinations probably diversifying, but only to a limited extent, from Europe to Asia. It is estimated that the energy sector as a whole (including exploration, development and maintenance of nuclear, coal, heat, electricity, gas and oil) will require around €715,000 million between 2003-2020.[79] Such massive investment will depend heavily on the effective openness of the overall Russian economy, and on the trust of foreign investors.

Much expectation surrounds the potential of the **Caspian region**.[80] Current oil reserves in the region are estimated between 17 and 44 billion barrels (more than half of these reserves are in Kazakhstan), and gas reserves are estimated at 6.4 trillion cm (most of these in Turkmenistan and Kazakhstan).[81] Many of these resources have not been tapped yet, and further reserves are suspected, but not proven.[82] However, 'future available infrastructure remains the key constraint to the expansion of exports from this region'.[83] Some prudent estimations envisage a peak in the Caspian oil production as early as 2010, at around 2.5-2.6 mb/d.[84]

Sub-Saharan Africa

Some African countries will hugely expand their production capacity. The oil production of Angola is expected to increase from 1.25 mb/d in 2005[85] to 3.2 mb/d in 2020. According to OPEC, the output of non-OPEC African producers could reach 8 mb/d in 2025, the bulk of the increase stemming from offshore fields in West Africa.[86] African proven oil reserves amount to 114 billion barrels (around 10% of the world oil reserves) but it is believed that more is yet to be discovered.[87] African oil exports already represent 28% of Chinese imports and 18% of American imports. Importantly, the latter could increase to 25% by 2015.

The aggregated African oil production (OPEC and non-OPEC) could comfortably match the foreseen continental demand (around 5 mb/d in 2030) but poor intra-continental transport infrastructure and pressing external demand make this prospect remote. Thus, the access to energy will remain one of the main constraints to development. Moreover, the increase in oil prices is likely to impact negatively on the growth of non-producer countries.[88]

Major energy importers

The European Union

While energy demand in the EU will continue to grow (by about +15%) by 2030, it will do so at a slower rate than in the past (and less than in the US).[89] Dependence on energy imports will grow (from current 50% to 70% in 2030),[90] with 90% of oil and 80% of gas being imported in 2025.[91] OPEC, and in particular Saudi Arabia, Iran, Iraq and Algeria, are expected to continue to provide around half of the EU's oil needs (currently 45%), while the remainder will come from Russia and Norway.[92] Gas imports are expected to come mainly from Russia (60%), followed by Norway and Algeria.[93] However, the growing use of LNG could help to diversify gas imports, with countries such as Qatar and Egypt as suppliers.[94]

The rapid increase in energy prices is reshaping the envisaged European energy mix. The expected share of oil in total energy consumption may decrease to 33.8% in 2030 (38.4% in 2000). Most remarkably, the respective trends of gas and coal consumption – growing for gas, and declining for coal – might be reversed between 2020 and 2030. The European Commission envisages that, after peaking in 2020, the share of gas will slightly decrease (from 28.1% in 2020 to 27.3% in 2030) whereas reliance on coal will increase (from 13.8% to 15.5%). The share of renewables should reach 12.2% in 2030 (5.8% in 2000). As far as nuclear energy is concerned, current projections still show an expected decrease (from 14.4% in 2000 to 11.1% in 2030).[95]

In 2004, power generation in the EU relied mainly on nuclear energy (31%) and coal (29%), with gas (19%), renewables (15%) and oil (4%) playing a less important role.[96] Electricity production is expected to grow by 51% up to 2030. Most of this increase should be covered by gas facilities, with a corresponding doubling of the share of gas in power generation. Renewable energy may account for 28% of the total power generation in 2030, and the share of nuclear power is expected to decrease to 19%, despite the renewed interest of some Member States and of the European Commission for this source of energy.[97] €625 billion will have to be invested to meet the energy demand, half of it to replace existing infrastructures and most of the rest to develop renewables.

The growing dependency of the EU on energy imports poses not only technical challenges but also and above all political ones. European institutions have raised the question of over-dependency on a few suppliers in recent key documents and have sketched out policy guidelines to avert the risk of energy distress.[98] Partnership and dialogue with suppliers and transit countries, the setting up of an international framework between producers and importing countries and the development of complementary or redundant infrastructures are some of the envisaged solutions.

The United States

Energy demand will increase faster in the United States than in any other OECD country (+36%, from 2002 to 2025).[99] Consumption of oil is expected to increase by 39% (from 19.7 mb/d in 2002), and consumption of natural gas by 34% (from 23.0 trillion cubic feet [cf] in 2002).[100] As with all mature economies, the US will need to increase oil and gas imports.[101] By 2030, it is expected that America will need to import 66% of its oil (47% in 2004) and around 20% of its gas (4% today) from abroad.[102] In addition, given decreasing gas production in Mexico and Canada, US demand for LNG might multiply by seven, with a huge impact on an already strained market.

However, compared to other energy-importing countries, the US has the strategic advantage of having important oil and gas reserves on its soil (amounting to 21.3 billion barrels and 602.8 trillion cubic feet respectively)[103] and a more diversified supply base (currently 33% of oil imports come from Latin America, 23% from the Middle East, 18% from Africa and 16% from Canada).[104]

President Bush has stated that the US aims at reducing its dependency on Middle Eastern oil further, replacing more than 75% of their imports from that region by 2025.[105] To achieve this objective, the US will foster renewable energies (current projections foresee a share of 14% of alternative energy sources in 2025, whereas nuclear energy is expected to stagnate at around 9-10%).[106] At the same time, the US is trying to further diversify its imports. The strategic importance of Africa and Latin America for US supplies will only grow. Africa is expected to account for 25% of US oil imports in 2025. Nigeria, already providing 10% of US oil supplies and expected to become a large source of LNG, will continue to play a pivotal role. Current attempts to enhance oil imports from Africa, however, bring US companies increasingly in competition with already established investors, such as BP and Total (in Nigeria, Congo, Angola and Gulf of Guinea), but also Chinese firms (already present in Sudan and Angola and intensively prospecting in other parts of Africa).

China

China's energy demand, already accounting for 12% of the world total, will boom over the next few decades.[107] In the period from 2002 to 2030, oil demand is expected to grow by 150% (from 5.3 to 13.3 million b/d), demand for gas by 336% (from 36 billion cm to 157 billion cm),[108] and demand for coal by 83% (from 1,308 Mt to 2,402 Mt).[109] Whereas coal demand can be satisfied from domestic reserves, oil and gas will have to be imported.[110] In this context, China is trying to reduce its dependency on the Middle East and Africa (who represent more than 50% of oil imports today), via direct investments in particular in Iran, Sudan, the Caspian region and Australia, and plans for a new oil pipeline connection with Russia.[111] China is also cooperating in the development of port facilities in Pakistan and Burma/Myanmar, which will provide platforms for alternative energy routes and for the projection of its naval power in the Indian Ocean.

To sustain current economic growth rates, China would have to increase its electricity production from 360 GW in 2002 to 1,220 GW in 2030.[112] Coal-fired power plants will remain the main source of energy production (776 GW planned in 2030), followed by gas-fired plants (111 GW). The remaining 333 GW are supposed to be produced by hydropower (200-240 GW), other renewables (38 GW) and nuclear energy (35 GW). However, many experts consider these projections as unrealistic (in particular for hydropower and renewables). To reach the objective of 1,220 GW, China would have to invest $2 trillion until 2030 only for power generation, which would represent 12% of the total world energy investments. Hydropower projections would imply the construction of a new Three Gorges Dam every two years.[113]

Energy supply in general, and power generation in particular, might therefore well slow down China's economic expansion. In any case, its growing demand will increasingly have an impact on energy prices. It may also reshape political alliances to control energy supply, with Chinese investments in Central Asia, Africa and the Middle East challenging the positions of the US, European countries and Russia.

India

India's energy demand is projected to increase by 109% between 2006 and 2025.[114] Oil consumption will increase by 123% (from 2.6 million b/d in 2004), and gas by 292% (from 28 billion cm in 2002). Coal consumption will grow considerably as well (by 94%, from 391 Mt in 2002), but be covered almost completely by domestic production (10% of world coal reserves are located in India).[115] By 2030, coal will still generate 64% of India's electric consumption, as opposed to 68% today. Nuclear energy growth rates will be high (+164%), with a production capacity of 1,346 billion kWh being planned by 2025.[116] It is envisaged that 8 nuclear reactors will be added to the 15 currently in operation.

Major investments in India's energy infrastructure are necessary to fulfil the projected needs. India's electric power demand is expected to more than triplicate by 2025, which will require the tripling of the installed generation capacity from 124 GW in 2005[117] to about 300 GW. In power generation alone, annual investment needs equate up to 2% of GDP until 2030. Since most of this investment would have to come from abroad, the fulfilment of India's energy (and economic) objectives will depend highly on the country's capacity to attract foreign (private) investors.

In any case, India will be very much dependent on foreign energy supply. Oil accounts for one third of India's total energy consumption. Domestic oil production, which covered half of consumption in the 1980s, will only satisfy 10% of the demand ten years from now.[118] About 62% of Indian oil imports come from four countries: Saudi Arabia, Kuwait, Iran and Nigeria. Discovery of new reserves could allow India to cover almost half of its expected gas demand. For the rest, political and geographical constraints are pushing India to invest in LNG, with several projects which by 2030 could cover 27% of the national demand. Most of the LNG imports will probably originate from the Middle East (notably from Qatar, Oman and Yemen, but also from Australia and Malaysia). In particular, India has signed a massive gas deal with Iran in 2005 to begin LNG imports in 2007, for a period of 25 years, aside from the envisaged pipeline connecting Iran to India via Pakistan (a project severely opposed by the US). Gas imports will also flow from Burma/Myanmar, where Indian firms have substantially invested in newly discovered reserves, and probably from Bangladesh, with proven gas reserves of 14 trillion cubic feet.

In addition to diversifying deals with potential suppliers, India is in the process of deploying a multi-pronged domestic strategy to address its vulnerability in the energy domain, including three main elements. First, India envisages intensifying the development of the nuclear share in power generation. Ultimately, by 2050, the nuclear sector is expected to produce roughly 20% to 25% of the country's power generation. However, given the lack of national reserves of uranium, India will be confronted with a dilemma between relying on the international market to buy uranium, or investing in an alterative option, based on fuel breeding and reprocessing and, in the medium term, on the thorium fuel cycle.[119]

Second, some envisage domestic alternatives to imported fossil fuels, namely the massive development of energy plantations to draw energy from biomass, as well as sustained investment in the production of biofuels, following the example of Brazil. The vast surface of degraded wasteland available in the country offers a unique opportunity for India.[120] Third, India is pursuing a more proactive policy of exploration of new oil and gas fields in its own territory, encouraging Western companies to intervene, and plans new investments to increase oil recovery rates from existing reserves and to improve the fuel efficiency of coal power plants.

Notes

1. Where possible, this section is based on data from the International Energy Agency's *World Energy Outlook (WEO)* (Paris, 2005). See also *World Energy Investment Outlook (WEIO)* (IEA, Paris, 2003); *Energy Security: Responding to the Challenge*, Brookings Briefing, 29 November 2005 (The Brookings Institution, Washington D.C.); *Dawn of a New Age: Global Energy Scenarios*, CERA (Cambridge Energy Research Associates) conference, 6 December 2005, Cambridge, Massachusetts.
2. IEA, *Findings of Recent IEA Work* (OECD/IEA, Paris, 2005), p. 8.
3. See IEA, *World Energy Outlook 2005*, op. cit. See also: 'Réflexions sur la dynamique des marchés du pétrole et du gaz naturel', *Déclaration 2004 du CME* (Conseil Mondial de l'Energie); *Review of long-term energy scenarios: implications for nuclear energy*, Moscow, Russian Academy of Science, April 2002, Power Point presentation; *Oil Outlook to 2025*, OPEC Review Paper, 2004.
4. *The World Energy Book*, World Energy Council, Issue 1, Autumn 2005, London, p.2 and see *Findings of Recent IEA Work*, 2005, op. cit
5. *Energy, Electricity and Nuclear Power Estimates for the Period up to 2030* (International Atomic Energy Agency [IAEA], July 2005 Edition, Vienna); 'Energy Prospects after the Petroleum Age', *Current Issues*, 2 December 2004, Deutsche Bank Research, Frankfurt. The latter quotes World Energy Council, *Energie Für Deutschland*, 2004; Khatib Hisham, *Energy Considerations – Global Warming Perspectives*, Energy Permanent Monitoring Panel of the World Federation of Science, 19 August 2004.
6. The renewable portfolio demand (hydropower, biomass combustion and geothermal technologies) is expected to increase by approximately 53% (from the current 32.1 quadrillion British Thermal Units). However, it will still represent a small portion of overall energy demand. See: IEA, *World Energy Outlook 2005*, op. cit.; Energy Information Administration, Official Energy Statistics from the US Government, available at: http://www.eia.doe.gov/oiaf/forecasting.html; Daniel Yergin, 'Questions of Oil', in: *The Economist: The World in 2006*, pp. 127-128.
7. *World Energy Outlook 2005* (IEA, 2005); *Middle East and Africa Insights* (OECD/IEA, Paris, 2005).
8. e.g. ExxonMobil, *The Outlook for Energy: A 2030 View*, 2005 edition.
9. The IEA has also produced the World Alternative Scenario, which takes into account envisaged but not yet adopted energy-efficiency and environmental measures. Under this scenario, overall energy demand would be 10% lower than in the Reference scenario. The demand for gas and oil would be lower by 10%, while the demand for coal would be lower by 23%. On the other hand, the use of renewables would be 27% higher. See IEA, *World Energy Outlook 2005*, op. cit.
10. See Deutsche Bank, 'Energy Prospects after the Petroleum Age', op. cit. and Christophe-Alexandre Paillard, 'L'influence des prix du pétrole sur l'économie mondiale', *Questions internationales*, no. 18, La Documentation française, Paris, March/April 2006.
11. See *The World Energy Book*, op. cit.
12. 'Quantifying Energy', *BP Statistical Review of World Energy*, June 2006.
13. See IEA, *World Energy Outlook 2005*, op. cit. The US Department of Energy gives higher figures. From 2003 to 2030 the increase is expected to be by 47%. From this increase, 43% may come from non-OECD Asia. *International Energy Outlook 2006*, DOE/EIA-

14. See IEA, *World Energy Outlook 2005*, op. cit.; *Oil Outlook to 2025*, op. cit.
15. Alan Larson, 'Geopolitics of Oil and Natural Gas', in: *Economic Perspective*, vol. 9, no. 2, May 2004.
16. Werner Zittel and Jörg Schindler, 'The Countdown for the Peak of Oil Production Has Begun – But what are the Views of the Most Important International Energy Agencies?', published on 14 Oct 2004 by *EnergyBulletin.net / EnergieKrise.de*.
17. *Oil Outlook to 2025*, OPEC Secretariat Paper, 10th International Energy Forum, Doha, 22-24 April 2006. There are, however, very pessimistic analyses of the prospects in OPEC countries themselves, and notably in Saudi Arabia. See Matthew R. Simmons, *Twilight in the Desert: The Coming Saudi Oil Shock and the World Economy* (John Wiley & Sons Inc., NJ, 2005); W. Zittel and J. Schindler, 'The Countdown for the Peak of Oil Production Has Begun', op. cit.
18. IEA, *World Energy Outlook 2004*, op. cit, and Warren R. True, 'WGC: Growth in gas trade faces range of challenges', *Oil and Gas Journal*, 12 June 2006.
19. e.g. Deutsche Bank, 'Energy Prospects after the Petroleum Age', op. cit. See also: e.g. IEA, *World Energy Outlook 2004*, op. cit. and ExxonMobil, 2005, op. cit., Warren R. True, 'WGC: Growth in gas Trade...', op. cit.
20. IEA, *World Energy Outlook 2005*, op. cit.
21. Ibid; World Energy Council, *The World Energy Book*, 2005, p. 2. See also Energy Information Administration, Official Energy Statistics from the US Government, available at: http://www.eia.doe.gov/oiaf/forecasting.html.
22. *Financial Times*, 16 January 2006, cited by the French Ministry of Environment and Sustainable Development (http://www.ecologie.gouv.fr/article.php3?id_article=5059).
23. 'Climate change, energy and sustainable development: How to tame King Coal?', Vision Paper, Coal Working Group, *Le Délégué Interministeriel au Développement Durable*, France, 9 June 2005. Revised version, 12 January 2006.
24. Today, coal provides 77% of China's electricity and 70% of India's. Coal also provides more than half of the energy needs of countries such as the US, Germany and Denmark. 'Climate change, energy and sustainable development: How to tame King Coal?', op. cit.
25. IEA, *World Energy Outlook 2005*, op. cit.
26. See IEA, *World Energy Outlook 2004*, op. cit., and ExxonMobil, 2005, op. cit.; also Fred Guterl, 'Another Nuclear Dawn', *Newsweek*, 6 February 2006, pp. 36-42.
27. EIA, *International Energy Outlook 2006*, op. cit.
28. See *A European Strategy for Sustainable, Competitive and Secure Energy*, Green Paper, European Commission, COM(2006) 105 final, 8 March 2006. See also C. A. Paillard, 'Strategies for Energy: Which way forward for Europe?', Fondation Robert Schuman, February 2006.
29. The nuclear sector currently represents a non-emission of around 180 million tons of CO_2. See 'Nuclear energy: can we do without it?', European Commission, *RTD Info*, no. 40, February 2004.
30. The Energy Challenge Energy Review Report 2006, Department of Trade and Industry, July 2006.
31. *How Much Bioenergy can Europe Produce Without Harming the Environment?*, EEA Report no. 7/2006, European Environment Agency, 2006. Nevertheless, massive reorientation towards biomass as a renewable source of energy could be difficult to sustain. For instance, to achieve the goals of the European Commission for Biofuels for 2010 (An EU Strategy for Biofuels, COM(2006) 34 final, Brussels, 8.2.2006) 'the extra crops required to bring

about the required *increase* in biofuel production (assuming 5.75% replacement of diesel by bio-diesel and 5.75% of gasoline by ethanol on an energy basis) would replace 27% of projected EU 2012 cereals production, or roughly 22% of total arable capacity (not including set-asides) or roughly 19% of arable capacity including set-asides', *Well-to-Wheels analysis of future automotive fuels and powertrains in the European context*, Well-to-Wheels Report, Version 2b, EUCAR, CONCAWE & JRC, May 2006.

32. IEA, *World Energy Outlook 2005*, op. cit.
33. e.g., ExxonMobil, 2005, op. cit.; IEA, *World Energy Outlook 2005*, op. cit.
34. IEA, *World Energy Outlook 2004*, op. cit.
35. Ibid.
36. IEA, *World Energy Outlook 2005*, op. cit., p. 89; IEA, *World Energy Outlook 2004*, op. cit.
37. IEA, *World Energy Outlook 2005*, op. cit.
38. *Annual Energy Outlook 2006*, DOE/EIA-0383(2006), February 2006.
39. *A European Strategy for Sustainable, Competitive and Secure Energy: What is at Stake*, Background document, Commission staff working document, Annex to the Green Paper, {COM(2006) 105 final}, SEC(2006) 317/2, Brussels, 2006; *Doing More with Less*, European Commission, Green Paper on Energy Efficiency, Brussels, 22 June 2005.
40. *Annual Energy Outlook 2006*, DOE/EIA-0383(2006), February 2006.
41. Indicative figures calculated by dividing the Chinese demand and the national production. Corresponding figures can be found in *WEO 2004*.
42. Ibid.
43. IEA. *World Energy Investment Outlook 2003*, op. cit.
44. The expected share of global international oil and gas maritime trade is constantly rising. Natural chokepoints such as the straits of Malacca and Hormuz will be intensively exploited: the IEA forecasts that gas and oil volume transiting through these two locations is likely to at least triple. See *WEO 2005* (IEA, 2005). The existence of such chokepoints and the potential strains in maritime capacity could incite some energy-dependent countries to invest in pipelines. China is thus considering the building of several pipelines to alleviate its maritime dependency. On the other hand, producer countries such as Russia may prefer to develop direct access to sea lanes, so as to ensure oil and gas revenues and to avoid dependency on particular clients, given the inflexibility of physical infrastructures overland.
45. These figures concern the extension of transmission pipelines in 2000. Distribution pipelines will be extended from 2 million km in 2000 to 8.5 million km in 2030. *WEO 2003* (IEA, 2003).
46. *World Energy Investment Outlook 2003*, op. cit.
47. Access to electricity in many parts of Africa is less than 200 kWh per person annually. In some African countries, the figure drops to under 30 kWh/person/year. As a comparison, access levels in the United States are over 12,000 kWh/person/year. See: 'African Regional Document: Water Resources Development in Africa', 4th World Water Forum, 16-22 March 2006. Accessible at http://www.worldwaterforum4.org.mx/home/tools.asp?lan=. See also: *The Energy Challenge for Achieving the Millennium Development Goals* (United Nations, UN-Energy, 2005); Jamal Saghir, *Energy and Poverty: Myths, Links and Policy Issues*, Energy Working Notes, Energy and Mining Sector Board, The World Bank Group, no. 4, May 2005.
48. Resource nationalism is defined as the policy of energy-producing nations directed to exclude or limit the foreign participation in the development of energy resources, and/or provide advantages to national companies in developing those resources. See Vladimir

Milov (ed.), *Deepening the integration between energy producing and consuming nations*, Institute of Energy Policy (Moscow), 25 January 2006.

49. See IEA, *World Energy Outlook 2005*, op. cit., p. 89. EUROGULF: An EU-GCC Dialogue for Energy Stability and Sustainability, Final Research Report (as presented at the concluding Conference in Kuwait, 2-3 April 2005), Project Ref.: 4.1041/D/02-008-S07 21089, 2005.
50. IEA, *World Energy Outlook 2005*, p. 91. It should be noted that some uncertainties exist on the validity of these projections, since the increase essentially depends on the Saudis' capacity to increase their own production from 10.5 mb/d in 2005 to 18.2 mb/d in 2030.
51. IEA, *World Energy Outlook 2005*, op. cit., p. 91.
52. Ibid., op. cit., p. 485.
53. Ibid., op. cit., p. 520.
54. Ibid., p. 521. For an alternative analysis of Saudi reserves and production capabilities, see Matthew R. Simmons, *Twilight in the Desert*, op. cit.
55. IEA, *World Energy Outlook 2005*, op. cit., p. 521.
56. 'Quantifying Energy', *BP Statistical Review of World Energy*, June 2006; *Annual Statistical Bulletin*, OPEC, 2004. According to OPEC figures, Iraq possesses the third highest reserves, and according to the IEA, the world's fourth highest reserves. See: IEA, *World Energy Outlook 2005*, op. cit.
57. See *The World Energy Book*, 2005, op. cit., p. 4.
58. IEA, *World Energy Outlook 2005*, op. cit., p. 373 and Jan Leijonhielm and Robert L. Larsson, *Russia's Strategic Commodities: Energy and Metals as Security Levers*, Swedish Defence Research Agency, November 2004.
59. IEA, *World Energy Outlook 2005*, p. 399.
60. Ibid., p. 357.
61. Ibid., p. 363.
62. Ibid., p. 364.
63. Ibid.
64. A new pipeline to Pakistan and India has been discussed. Moreover, Tehran is thinking about transporting its gas to Europe via Bulgaria, Romania and Hungary (the Nabucco Pipeline) and has already signed an agreement to extend the Iranian-Turkish pipeline to northern Greece. See: IEA, *World Energy Outlook 2005*, op. cit., p. 368.
65. See IEA, *World Energy Outlook 2005*, op. cit.
66. Energy Information Administration, Official Energy Statistics from the US Government, available at: http://www.eia.doe.gov/oiaf/forecasting.html.
67. Energy Information Administration, Official Energy Statistics from the US Government, available at: http://www.eia.doe.gov/oiaf/forecasting.html.
68. *Annual Statistical Bulletin*, OPEC 2004 and EIA Country analysis for Nigeria (http://www.eia.doe.gov/).
69. EIA Country analysis for Libya (http://www.eia.doe.gov/).
70. Energy Information Administration, Official Energy Statistics from the US Government, available at: http://www.eia.doe.gov/oiaf/forecasting.html and e.g., IEA, *World Energy Outlook 2005*, op. cit.
71. *Oil Outlook to 2025*, OPEC, 2004, op. cit.
72. Leijonhielm and Larsson, 2004, op. cit.
73. IEA, *World Energy Outlook 2004*, op. cit.

74. *Oil Outlook to 2025*, OPEC, 2004, op. cit.
75. Statement of Sergeï Oganesyan (Director of the Russian Energy Agency) in 'Russia: Oil production to be flat in 2005', *Associated Press*, Moscow, 4 June 2004.
76. *Rising Energy Competition and Energy Security in Northeast Asia: Issues for US Policy*, Congressional Research Service (CRS), 14 July 2004.
77. http://eng.gazpromquestions.ru/page7.shtml. See also: Marina Kim, 'Russian Oil and Gas: Impacts on Global Supplies to 2020', *Australian Commodities*, vol. 12, no. 2, June Quarter 2005 and see IEA, *World Energy Outlook 2005*, op. cit.
78. See IEA, *World Energy Outlook 2004*, op. cit.
79. Christian Cleuntinx, 'The EU-Russia Energy Dialogue' (DG for Energy and Transport, European Commission, Vienna, December 2003).
80. See Alan Larson 2004, op. cit; *Caspian Oil and Gas: Production and Prospects*, Congressional Research Service (CRS), 4 March 2005.
81. Energy Information Administration (EIA), Official Energy Statistics from the US Government, http://www.eia.doe.gov/emeu/cabs/Caspian/Background.html; Dov Lynch (ed.), 'The South Caucasus: a challenge for the EU', *Chaillot Paper* no. 65 (EUISS, Paris, 2003).
82. *British Petroleum Statistical Review of World Energy*, 2002.
83. *Oil Outlook to 2025*, OPEC, 2004, op. cit.
84. W. Zittel and J. Schindler, 'The Countdown for the Peak of Oil Production Has Begun...', op. cit. By 2010, OPEC expects the Caspian area and other former Soviet Union producers (Russia excluded) to produce around 3.9 million barrels a day and 5.3 mb/d in 2025; *Oil Outlook to 2025*, OPEC Secretariat Paper, op. cit.
85. EIA, *Country Analysis: Angola*, available online on http://www.eia.doe.gov/.
86. *Oil Outlook to 2025*, OPEC, op. cit.
87. This evaluation, however, drops to 63 billion barrels if Libya and Algeria are excluded from the statistics. See 'Quantifying Energy', *BP Statistical Review of World Energy*, June 2006.
88. It should be noted, however, that the effect on the poorest countries could be less severe than foreseen since biomass represents the main energy source under $300 per capita GDP (in PPP terms).
89. *European Energy and Transport Trends to 2030* (DG for Energy and Transport, European Commission, January 2003).
90. Robert Willenborg et al., *Europe's Oil Defences. An Analysis of Europe's oil supply vulnerability and its emergency oil stockholding systems* (The Clingendael Institute, The Hague, 2004), p. 15. 'Unless consumption rates show a downward trend in the most rapidly growing sectors – transport and housing – Europe's energy dependence will reach more and more worrying levels'. EU Green Paper, *Towards a European Strategy for the Security of Energy Supply*, European Commission Green Paper (COM(2000) 769 final), 29 November 2000.
91. *Doing More with Less*, Green Paper on Energy Efficiency, European Commission, 22 June 2005, Brussels, p. 5; EUROGULF: An EU-GCC *Dialogue for Energy Stability and Sustainability*, Final Research Report (as presented at the Concluding Conference in Kuwait, 2-3 April 2005), Project Ref.: 4.1041/D/02-008-S07 21089, 2005.
92. According to the Green Paper, *Towards a European Strategy...*, op. cit., OPEC will cover 50% of EU needs by 2020 (with 94% of this coming from MENA countries). The more recent DOE/EIA *International Energy Outlook*, op.cit., sets the share of EU oil imports from MENA in 2025 at around 45%.
93. See: EU Green Paper, *Towards a European Strategy...*, op. cit. and EurActiv, *Géopolitique des*

approvisionnements énergétiques de l'UE, op. cit.

94. See Jörn Sucher, 'Europa zapft jetzt Flüssiggas' , *Spiegel-online*, 23 March 2006.
95. *A European Strategy for Sustainable, Competitive and Secure Energy: What is at Stake*, op. cit.
96. Eurostat. *Energy: Yearly Statistics. Data 2004*, June 2006 (http://epp.eurostat.ec.europa.eu/cache/ITY_OFFPUB/KS-CN-06-001/EN/KS-CN-06-001-EN.PDF).
97. 'The EU can play a useful role in ensuring that all costs, advantages and drawbacks of nuclear power are identified for a well-informed, objective and transparent debate', *A European Strategy for Sustainable, Competitive and Secure Energy*, Green Paper, European Commission, [SEC(2006) 317], COM(2006) 105 final, Brussels, 8 March 2006.
98. In envisaging a future policy on securing and diversifying energy supplies, the Commission argued that 'Such a policy is necessary both for the EU as a whole and for specific Member States or regions, and is especially appropriate for gas. To this end, the above-mentioned Review could propose clearly identified priorities for the upgrading and construction of new infrastructure necessary for the security of EU energy supplies, notably new gas and oil pipelines and liquefied natural gas (LNG) terminals as well as the application of transit and third party access to existing pipelines. Examples include independent gas pipeline supplies from the Caspian region, North Africa and the Middle East into the heart of the EU, new LNG terminals serving markets that are presently characterised by a lack of competition between gas suppliers, and Central European oil pipelines aiming at facilitating Caspian oil supplies to the EU through Ukraine, Romania and Bulgaria'. *A European Strategy for Sustainable, Competitive and Secure Energy*, Green Paper 2006, op. cit. See also the subsequent Paper from the Commission and the Secretary General/High Representative for CFSP, submitted to the European Council in June 2006, *An External Policy to Serve Europe's Energy Interests*.
99. Energy Information Administration, Official Energy Statistics from the US Government, available at: http://www.eia.doe.gov/oiaf/forecasting.html.
100. Ibid.
101. e.g. *Energy Security: Responding to the Challenge*, op. cit.
102. *Annual Energy Outlook 2006*, DOE/EIA-0383(2006), February 2006.
103. e.g. Deutsche Bank, 2004, op. cit.
104. Energy Information Administration (http://www.eia.doe.gov/).
105. State of the Union Address by President George W. Bush, United States Capitol, Washington D.C., 2006.
106. 'World Total Energy Consumption by Region and Fuel, Reference Case, 1990-2025', Energy Information Administration, http://www.eia.doe.gov/oiaf/ieo/pdf/ieoreftab_2.pdf.
107. ExxonMobil, 2005, op. cit. and also *Evaluation of China's Energy Strategy Options*, The China Sustainable Energy Program, China Energy Group, Berkeley, CA, 2005; see also IEA, *World Energy Outlook 2005*, op. cit.
108. Energy Information Administration, Official Energy Statistics from the US Government, available at: http://www.eia.doe.gov/oiaf/forecasting.html.
109. See IEA, *World Energy Outlook 2005*, op. cit., p. 87 and Energy Information Administration, Official Energy Statistics from the US Government, available at: http://www.eia.doe.gov/oiaf/forecasting.html.
110. 'Climate change, energy and sustainable development: How to tame King Coal?,' op. cit., p. 21 and Energy Information Administration, Official Energy Statistics from the US Government, available at: http://www.eia.doe.gov/oiaf/forecasting.html.
111. *Rising Energy Competition and Energy Security in Northeast Asia: Issues for US policy*, op. cit. More specifically, according to the US Department of Energy, 'the Chinese have significantly

increased the number and geographic distribution of energy assets and investments, although total overseas oil investments by Chinese firms remain small compared to investments by the international oil majors. 56 Chinese national oil companies have invested in oil ventures in over 20 countries with bids for oilfield development contracts, pipeline contracts, and refinery projects in Iran, Sudan, Kazakhstan, Kuwait and others. In addition, the Chinese have recently focused on broadening their equity stakes in North Africa, Central Asia, Southeast Asia, Latin America and most recently in North America, where they have acquired stakes in Canadian oil sands firms and unsuccessfully attempted to acquire the US firm Unocal.' Energy Policy Act 2005, Section 1837: National Security Review of International Energy Requirements Prepared by The US Department of Energy, February 2006.

112. IEA, *World Energy Outlook 2004*, op. cit.
113. Ibid.; see also Jonathan E. Sinton, Rachel E. Stern, Nathaniel T. Aden and Mark D. Levine, *Evaluation of China's Energy Strategy Options*, Report prepared for and with the support of the China Sustainable Energy Program,16 May 2005; *World Energy Investment Outlook – 2003 Insights* (IEA, Paris, 2003).
114. *India – A Growing International Oil and Gas Player* (IEA, Paris, 2000).
115. *World Energy Outlook 2004*, op. cit. See also Energy Information Administration, Official Energy Statistics from the US Government, available at: http://www.eia.doe.gov/oiaf/forecasting.html and e.g., IEA, *World Energy Outlook 2005*, op. cit.
116. Energy Information Administration, Official Energy Statistics from the US Government, available at: http://www.eia.doe.gov/oiaf/forecasting.html.
117. Ministry of Power, India, data from the Ministry's website, http://powermin.nic.in/JSP_SERVLETS/internal.jsp.
118. In 2003 India was already depending on oil imports by 70%; see IEA, *World Energy Outlook 2004*. Brahma Chellaney, 'India's Future Security Challenge: Energy Security', in *India as a New Global Leader* (The Foreign Policy Centre, 2005).
119. See chapter on India (pp.165-76).
120. Report to the Committee on India Vision 2020, Planning Commission, Government of India, New Delhi, 2002.

Environment

4

The negative environmental consequences of the current model of industrial development are widely recognised,[1] but a sustainable alternative is still a long way off. This is the case for developing countries in particular, but even OECD countries will, at best, be able to move only gradually to a more sustainable model through economic incentives and technological innovation.[2] Moreover, the possible reduction of pollution in developed countries will be more than offset by the increase in pollution in emerging economies (in particular in Asia). As a consequence, the global environment is expected to further deteriorate. However, given the complexity of ecosystems and the multitude of inter-related factors that impact on them, future developments and consequences are difficult to forecast and are often scientifically controversial. This is the case in particular for climate change and its interaction with pollution.

General Trends

- The emission of greenhouse gases is expected to continue to grow. Their dissolution in the atmosphere takes at least several decades.[3] Global warming is therefore a long-term trend, which in the foreseeable future can be mitigated but not stopped.

- By 2025, temperatures are expected to rise worldwide between 0.4°C and 1.1°C,[4] with important regional variations in precipitation. Over the next twenty years, this would probably not imply a dramatic climate change with catastrophic consequences.

- However, the impact of global warming will become increasingly noticeable in certain arid and semi-arid areas of Africa and Asia, where the environment has already been damaged and the effects of pollution are particularly visible.

> In the short to medium term, industrialisation and urbanisation will be the main sources of environmental degradation.[5] In developing countries in particular, they will increase pollution and put water supply, sanitation capacities and food security increasingly under stress.[6]

Controlling anthropogenic (man-made) greenhouse gases (GHG) is one of the main challenges for the world economy. GHG emission and holes in the ozone layer caused by human activities (industry, heating, transport) are presumed to be responsible for a global rise in temperatures. The level of carbon dioxide before the Industrial Revolution was around 280 parts per million by volume (ppmv) and is currently around 370 ppmv. The amount of emission is correlated to energy use and, on the basis of the expected rates of economic growth and envisaged environmental policies, levels of CO_2 are expected to rise up to between 490 and 1,260 ppmv by 2100.[7]

According to the IEA, the expected growth in energy consumption will imply a **1.6% growth per year in the emissions until 2030**.[8] Developed countries would thus increase their emission from 12.7 billion tonnes of CO_2 to 15.3 billion tonnes in 2030. By then, they will be overtaken by the developing economies, which may emit around 18.1 million tonnes of CO_2. Those figures, however, can only be indicative: several European governments, already committed to the Kyoto Protocol, intend to tighten their emission reduction policies after the 2012 deadline of the Protocol. The main emission producer, the US, is implementing a more stringent environmental policy. In parallel, emerging economies such as China and India intend to dramatically increase their energy efficiency.

Moreover, emerging economies (designated as Non-Annex 1 in the Kyoto Protocol, and hence not committed to cut their emissions) are involved in negotiations on establishing ceilings on the levels of emission. Nonetheless, curbing global warming would require greater determination. By 2100, worst-case scenarios forecast an atmospheric GHC concentration of around 1,200 ppmv. To effectively mitigate global warming, those concentrations would need to be brought down to a third of that level. Energy efficiency improvement is a precondition for slowing down emissions but strategic choices between fossils resources, nuclear power and renewables will have to be made promptly in developed as well as in developing economies.

Environment

Energy-related CO_2 emissions (million tonnes of CO_2)

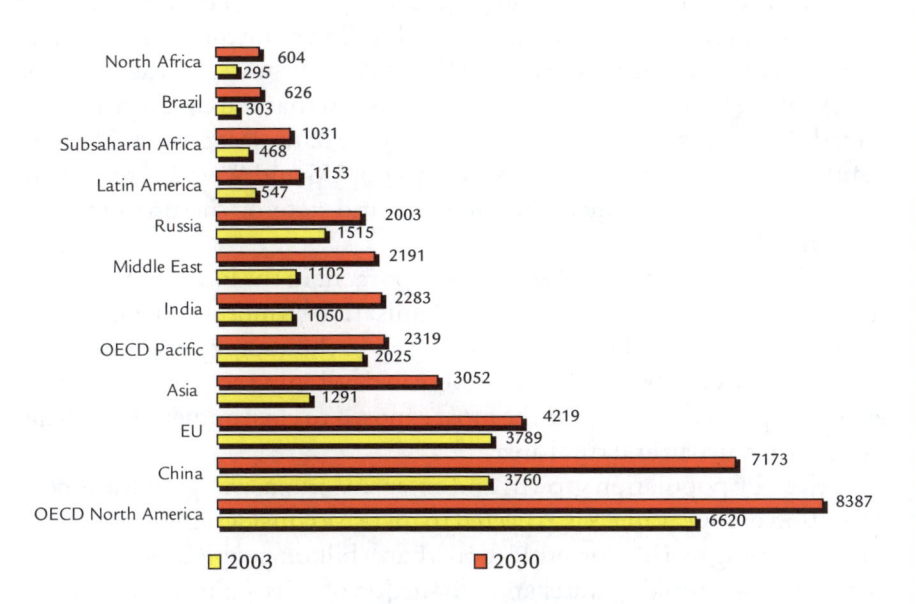

Region	2003	2030
North Africa	295	604
Brazil	303	626
Subsaharan Africa	468	1031
Latin America	547	1153
Russia	1515	2003
Middle East	1102	2191
India	1050	2283
OECD Pacific	2025	2319
Asia	1291	3052
EU	3789	4219
China	3760	7173
OECD North America	6620	8387

Source: International Energy Agency, *World Energy Outlook 2005*.

It follows that, regardless of specific policy initiatives, **the rise in temperature seems inescapable in the long term.** In arid and semi-arid areas (the Middle East and North Africa, the Horn of Africa, Southern Africa, North-West China, Central Asia), a rise in temperatures and a decrease in rainfall will affect river discharge, vegetation and soils, and cause droughts and desertification. The impact on agriculture and water resources will be severe, even in the short term. In temperate regions (Western Europe or Eastern parts of the United States, for instance) and tropical areas, temperatures will rise and rainfall will increase. This is likely to cause floods and heatwaves.[9] Tropical areas are also liable to be threatened by tropical storms, typhoons and hurricanes.[10]

The melting of the polar icecap and of the glaciers is likely to intensify global warming. A 2004 report suggests that the average temperature in the Arctic has risen at almost twice the rate of other parts of the world over the last few decades. The reduction of the sea ice area is now assessed to be at a rate of 8% per decade since 1978. The implications of Arctic warming are wide-ranging, spanning from rising sea levels to changes in global biodiversity.[11]

Sea levels are expected to rise only slightly (between 0.14 cm to 0.3 cm) and not to endanger costal areas by 2030.[12] If the current trend persists, however, the Arctic ice cap could disappear by 2060.[13] The desalinisation of Arctic water, due to ice melting, may slow down the thermal exchanges between hot and cold streams in the North Atlantic waters, weakening the effects of the Gulf Stream. Recent studies assess that the Gulf Stream may already have weakened by 30% since the middle of the last century.[14] A substantial alteration may in turn severely cool US and West European climates but the probability, the intensity and date of the phenomenon remain unknown.[15]

Global warming will change the environment in the long term, whereas industrialisation, over-farming and urbanisation[16] (more than population growth as such)[17] will have an impact even in the short to medium term. Consequences will be particularly severe in developing countries.[18] In general, poor populations will be the most exposed to environmental degradation, both in urban and rural areas.[19]

Driven by population growth and extensive/intensive agriculture, **consumption of blue water[20] is expected to increase considerably throughout the developing world**. Currently more than a billion people are deprived of access to clean drinking water and exhaustion of soils, pollution and desertification will diminish resources. This will put at risk blue water supply in particular in arid and semi-arid areas, where almost 90% of available resources will probably be exploited by agriculture in 2030.[21] Management of trans-boundary water resources could thus become a critical issue.[22]

The decimation of rain forests through extensive agriculture, and consumption of wood for heating,[23] is likely to continue, intensifying greenhouse gas emissions and soil erosion. Forests can function as carbon 'sinks', absorbing CO_2, or convert into a major source of emissions, when the woods are felled and burned.[24] In the 1990s, out of a total of 1,950 million hectares of forest, 15.2 million hectares were lost each year essentially because of changes in land use (roughly 8% of the total forest surface has disappeared over the last decade).[25] If temperate forests in Europe and the US are considered as carbon sinks, tropical forests have become net producers of carbon. Their progressive reduction could therefore be an important source of the release of greenhouse gases.

African, Asian and South American forests totalise 216 billion tons of carbon and figures for the 1980s quantify the release of carbon in tropical forests at 2-2.4 billion tonnes annually. Analysis of deforestation trends shows a slowdown in worldwide net losses, but not a fundamental modification of the deforestation process over the last 20 years.[26] Thus, a real policy turn would be necessary to restore the role of tropical forests as carbon sinks. By 2050, the restoration of damaged tropical forests and better forest management could allow a saving of between 11.5 and 28.7 billion tonnes of carbon,[27] although economic and demographic pressures may well thwart initiatives in this direction.[28] According to some scenarios, moreover, global warming could lead to a reduction of the boreal forest of up to 36%. This forest is one of the main reservoirs of carbon (74 billion tonnes in the vegetation and 249 billion tonnes in soils).

Driven by increasing industrialisation and urbanisation, **pollution will continue to increase**. The effects of industrial pollution on human health are well-known (e.g. allergies, poisoning of resources), but human pollution will become increasingly important, in particular in urban centres of developing countries. In these areas, pollution of waters by sewerage and dumping of rubbish leads to disease (e.g. cholera, diarrhoea) and is already a major factor of mortality, which will certainly get worse. Some experts estimate the amount of annual deaths caused by water-related diseases at up to 5 million. It is expected that respiratory ailments, caused by biomass fuel consumption (wood, coal, oil etc),[29] should affect about 3.7 billion people in 2030, in particular in the urban centres of developing countries.[30]

Regional differences

Europe

Though highly industrialised and urbanised (76% of the European population will live in cities in 2030),[31] Europe should not be too heavily affected by environmental degradation by 2025. The main risk to health is expected to come from air pollution, mainly due to industry and transport. Temperatures in Europe are expected to rise between 0.1°C and 0.4°C by decade. Mediterranean countries, on the one hand, and Finland and Western Russia, on the other, are likely to suffer the most acute variations in this regard. Precipitation is expected to decrease in Southern Europe (-1% per decade, -5% in summer), but to increase elsewhere (1-2% per decade).[32]

Climate warming should firstly affect water resources (including glaciers), mostly in Southern Europe where droughts will be more frequent and river discharge continue to decrease.[33] Droughts, heatwaves and water stress may also cause shifts in agricultural production.[34] Increased crop yields in Northern Europe will probably offset losses in Southern Europe, but European crop yield growth as a whole is likely to slow down. Most of Europe could suffer from some degree of water stress: the Southern part of it due to lack of supply, the North-Western part due to the increase in demand, and the Eastern part of the continent due to both.[35] Heatwaves[36] and coldwaves may affect the health of populations and also impact on infrastructures, in particular electricity networks. In Northern Europe, floods (caused by increased precipitation) are seen as a real but nevertheless manageable threat.

The EU is an Annex 1 Party of the Kyoto Protocol and is thus bound to reduce its greenhouse gas emissions to 8% below its 1990 ceiling by 2010.[37] Unlike the US, the EU has opted for a 'market pull' approach to limit its emissions: financial incentives are supposed to lead the market to upgrade its industrial capabilities. The EU Emissions Trading Scheme (ETS) has been established to trade carbon credits on the international market and to complete the Kyoto mechanism to reduce emissions. However the measures that have been undertaken appear insufficient: according to the European Commission, CO_2 emissions should 'increase significantly, exceeding the 1990 level by 3% in 2010 and by 5% in 2030'.[38] Stricter national policies will have to be implemented to respect the Kyoto targets and additional commitments in this direction are currently being taken by EU governments individually.

The United States

The US currently accounts for about 25% of worldwide emissions of greenhouse gases and is expected to reduce its share only slightly to 22% in 2030.[39] As in other parts of the world, global warming and its effects will increase throughout the twenty-first century,[40] but tangible signs are already expected to become visible in the run-up to 2020. The most likely phenomena will include increased dryness in Western parts of the country, higher rainfall in temperate regions, warmer winters in the North, but also migration of disease-carrying species, in particular from Florida to the North. Intensive agriculture and urbanisation in semi-arid areas are bound to put a heavy strain on water resources in the Southern and Western parts of the country.[41] The number of tropical storms and hurricanes in the Gulf of Florida, Gulf of Mexico and in the Caribbean is also expected to increase (although it is unclear whether this trend is linked to ocean warming).[42]

Concerning the environmental policy, the Administration has chosen to implement a national policy disconnected from the Kyoto commitments. The 'Energy Security Act 2005' does not refer to GHG.[43] Current environmental policy guidelines are based on an 18% cut in greenhouse gas 'intensity' from 2002 to 2012,[44] and a 70% decrease of sulphur dioxide, nitrogen oxide, and mercury by 2020. $4.1 billion of fiscal incentives is to be made available over the next five years for renewable energy and hybrid/fuel-cell vehicles, and $1 billion for the development, over a ten-year period, of a coal-based, zero-GHG emissions, power plant prototype.

Nonetheless, the priority given to economic growth and the lack of environmental control at the federal level explains the significant increase in GHG emissions from the US. Between 2002 and 2030, US emissions may increase by 29% whereas Europe plans to adopt more stringent measures to contain the increase in its emissions. On the other hand, should the US government revive a stronger nuclear programme, together with modernisation of the transport and the coal-generating infrastructure, its capacity to address the GHG issue could be significantly enhanced. In perspective, the US focus on technological innovations, as opposed to multilateral frameworks potentially slowing down economic growth, could lead to alternative policy options from those envisaged by the Kyoto Protocol, and shape a new (or parallel) environmental agenda.

Latin America

The number and intensity of violent natural disasters in Latin America is expected to increase (notably tropical storms in Central America), which could entail serious dangers for local populations, largely residing in coastal areas.[45]

The future of the Amazon forest, considered as one of the planet's main reservoirs of CO_2, remains bleak. The pace of deforestation, linked to infrastructural developments and extensive agriculture, should remain intense. Out of an overall extension of 7 million km², about 25,000 km² of the Amazon forest disappears each year (16% of the forest has already been lost).[46] It is estimated that, short of a determined policy change, 30% to 42% of the Amazon forest will be severely damaged by 2020, reinforcing climate change through massive greenhouse gas emissions.[47] The extension of human activities – mostly agriculture – at the expense of the Amazon forest is achieved partly through land burnings, thus releasing additional greenhouse gases. Those alterations could locally dry the climate of the Amazon basin. This modification could, in turn, accelerate the transformation of the rain forest into bushes and savannahs, increasing the effects of natural and human-induced forest fires and the release of greenhouse gases with a spillover effect.

Sub-Saharan Africa

Until 2025, global warming is expected to have a mixed impact in Africa. In certain regions, warming and changing precipitation patterns, in particular in combination with over-farming, are likely to lead to a noticeable deterioration in environmental conditions.[48] Some manifestations of climate change, like the quasi-disappearance of Lake Chad (which has shrunk from 25,000 km² to 1,200 km²) or the melting of the ice cap of the Kilimanjaro, illustrate the potential impact of global warming on the continent. Violent natural events (essentially droughts and floods) caused by global warming or by the El Niño Southern Oscillation (ENSO) could put parts of the population at risk. At the same time, Africa will be much more affected by human than by industrial pollution.

Africa's lack of access to water is particularly worrisome since water resources are abundant but very unevenly distributed; most of them are concentrated in Central Africa and, to a lesser extent, in West Africa. Currently, about 300 million people lack access to a suitable water supply and roughly 313 million people lack access to appropriate sanitation. Overall, fourteen countries currently suffer from either water stress or water scarcity.[49] By 2025, this number is expected to have almost doubled. Worst-case scenarios forecast 40% of African people suffering from water stress in 20 years from now.[50] In particular in the semi-arid areas of Southern Africa and the Horn of Africa, rising temperatures and decreasing precipitation should aggravate already existing water shortages and soil erosion (desertification currently affects about 46% of Africa). Even in equatorial Africa, where precipitation is expected to increase, human intervention (dams, pollution, deforestation) could lead to water shortage.[51]

In these regions, extensive agriculture, deforestation, and domestic wood consumption will enhance soil erosion and decrease crops output.[52] Those trends could worsen the deficiencies of the agriculture sector. Africa is already depending on cereal imports (currently 10 million tons per year) and 33% of the population is currently undernourished, with peaks of 50% in some countries. Projections assess that this dependency will deepen (30 million tons) unless agricultural output increases annually by 3.3%.

At the same time, warmer temperatures will accelerate the maturation of parasites in mosquitoes, which is likely to increase the spread of infectious diseases, such as malaria or dengue.[53] Due to the migration of infectious disease-carrying species, these diseases are likely to expand to temperate areas as well.

The Middle East and North Africa

The MENA region is the most arid in the world with more than 87% of its 14 million km² consisting of desert. In the future, it is expected to witness even higher temperatures and less rainfall. Combined with growing water consumption (due to a rapidly growing population, urbanisation, industrialisation, and extensive agriculture), this is likely to lead to a serious shortages in the water supply.⁵⁴ The international management of this resource will be imperative since major countries, such as Egypt and Syria, largely depend on freshwater resources (rivers) originating from other countries, and are therefore vulnerable to the water policies of their neighbours.⁵⁵ The management of water basins, however, is increasingly subject to multilateral arrangements in the Maghreb/Mashreq area (Nile Basin Initiative, Turkish-Syrian Joint Water Committee).

Under current estimates, annual per capita water availability, currently at about 1,200 m³, should shrink to 600/500m³ in 2025.⁵⁶ The sustainability of the human and economic development of the region will thus depend on long-distance water transfers and massive desalinisation projects.⁵⁷ Without appropriate and expensive measures, by 2050 the whole region (with the exception of Iraq) could experience water stress or water scarcity.⁵⁸ In the medium term, water scarcity will affect agriculture and increase the region's dependence on food imports, when 80 million tons of food are already imported each year. Rapid population growth and water scarcity in rural areas will lead to growing urbanisation. In 2015, 70% of the region's population is expected to live in cities.⁵⁹ This will put urban infrastructures under enormous pressure and may lead to considerable supply problems.

China

Pollution and environmental degradation cost China around 10% of its GDP.⁶⁰ Due to its reliance on fossil fuels, China's share of world CO_2 emissions is expected to rise from 15.2% to 19% in 2025.⁶¹ China is a non-Annex 1 party to the Kyoto Protocol and is not legally bound to reduce its emissions, even if China intends to control them through national programmes (energy efficiency, modernisation of transport, etc). Nonetheless, the effects of pollution are already tangible. Combined with biomass burning, polluting emissions already cause acid rain on 30% of China's territory, a quarter of which is also exposed to rapid desertification.⁶²

Over the last 10 years or so, China has lost almost 8 million hectares of farmland⁶³ and the process continues at the pace of 200,000-300,000 hectares per year. Although the decrease of arable lands is not exclusively caused by pollution (but is also due to restoration of the natural environment, which is, according to the Chinese Ministry of Environment, the main cause of loss of arable land), their deterioration is an additional cause for concern (affecting 37.1% of total arable surface according to official figures).⁶⁴ Some studies expect that 10 million hectares of arable land could be lost by 2030.⁶⁵ These developments have also put considerable strain on food supplies.⁶⁶ Turning to water, the picture looks no better, with 75% of major rivers polluted and 180 million people depending on contaminated water. Northern regions experience important shortages, notably in the valleys of the Yellow, Huaihe and Haihe rivers.[67] Water scarcity is such that official statements describe it as 'an unavoidable issue threatening national security'.[68] It has been estimated that environmental degradation may result in 20 to 30 million 'environmental refugees' up to 2020, adding to the huge number of economic migrants within the country.[69]

The magnitude of the coming environmental crisis is such that a massive intervention will be required to change energy production and consumption patterns, improve soil and water management and reform agriculture. Investments in water management have been evaluated at tens of billions of dollars and some huge projects, such as the north-south water diversion scheme,[70] have been launched. Nevertheless, ageing infrastructure, budgetary constraints and mismanagement will continue to pose major obstacles.

South Asia / India

Increasingly violent monsoons and typhoons are likely to strike coastal areas of Southern Asia and India. Consequences could be severe since the Asia Pacific region is already the most exposed region in the world (accounting for 91% of the worldwide death toll due to violent natural events and 49% of the concomitant worldwide economic damage).[71] Floods and a rise in sea levels could threaten low-lying lands (Bangladesh, the deltaic coasts of China and South East Asia) and impoverish soils.

Geographical constraints (mountainous relief and forests) tend to drive populations onto coastal land, notably in South East Asia. Those concentrations will facilitate the spread of diseases, increase vulnerability to violent natural events[72] and heighten the impact of pollution on health. Infectious/epidemic diseases linked to global warming and urbanisation are likely to spread and are bound to widely affect human health in South Asia.

Urbanisation and industrialisation have already caused a massive haze layer (atmospheric brown clouds) to appear for 3 or 4 months a year in the skies over parts of India and South Asia. Caused by anthropogenic aerosols (essentially resulting from biomass and fossil fuel burning), they have a different impact from greenhouse gases on local climate, agriculture and health.[73] Haze lowers temperatures and increases dryness, implying changes in the vegetation distribution. The impact of pollution on health is not precisely estimated. It is clear, however, that indoor pollution poses a particularly serious problem and respiratory infections are the second cause of death in South Asia.[74]

Notes

1. *Environmental Outlook 2001* (OECD, Paris, 2001); *Sustainable Use and Management of Natural Resources*, EEA Report 9/2005 (European Environment Agency, 2005); *Europe's Environment: The Third Assessment*, EEA Environmental Assessment Report no. 10 (European Environment Agency, 2003).
2. Kirsten Halsnæs and Priyadarshi R. Shukla, *Mainstreaming International Climate Agenda in Economic and Development Policies* (World Meteorological Organisation, 2005); *Analysis of Costs to Abate International Ozone-Depleting Substance Substitute Emissions* (US Environmental Protection Agency, EPA 430-R-04-006, 2004); *Deploying Climate-friendly Technologies through Collaboration with Developing Countries*, IEA Information Paper (International Energy Agency, 2005); *Using the Market for Cost-Effective Environmental Policy*, EEA Report 1/2006 (European Environment Agency, 2006); Stephen R. Connors, Warren W. Schenler, *Climate Change and Competition – On a Collision Course* (Massachusetts Institute of Technology, 1999).
3. Removal times for carbon dioxide and nitrous oxide are approximately 100 years, 10 years for methane, but 1,000 years for perfluorocarbon compounds. *Climate Change Science* (Committee on the Science of Climate Change, National Academy Press, Washington D.C., 2001).
4. Figures differ according to data included in climate models, and according to the model selected. The figures used in this report are the most commonly cited and come from the Intergovernmental Panel on Climate Change (IPCC) studies. See: *Climate Change 2001: Synthesis Report* (IPCC, 2001). On the IPCC, see Bernd Siebenhüner, *The Changing Role of Nation States in International Environmental Assessments: the case of the IPCC*, Global Governance Project Working Paper no. 7, July 2003.
5. The European Environment Agency identifies demography and industrialisation as the two main factors of environmental stress; see *Sustainable Use and Management of Natural Resources*, EEA Report 9/2005 (European Environment Agency, 2005); *Global Environmental Outlook 3* (United Nations Environment Programme, 2002).
6. Jill Boberg, *Liquid Assets: How Demographic Changes and Water Management Policies Affect Freshwater Resources* (MG-358, Rand Corporation, 2005).
7. Anthropogenic greenhouse gases include carbon dioxide (CO_2), methane (CH_4), nitrous oxide (N_2O) and tropospheric ozone (O_3). Intergovernmental Panel on Climate Change (IPCC), *Special Report on Emission Scenarios* (SRES), http://www.grida.no/climate/ipcc/emission.
8. *World Energy Outlook 2005* (International Energy Agency, Paris, 2005).
9. Natural cyclic phenomena, like the El Niño-Southern Oscillation (ENSO) and North Atlantic Oscillation (NAO), will regionally modify the impact of global warming. Aerosol emissions may locally modify those trends.
10. Data suggest that approximately 90% of natural disasters in recent times are weather and climate- related. Statistics also show a net increase of violent natural phenomena but the precise link with global warming has not yet been established. Cyclical trends, caused by ENSO for instance, can also be at the origin of some of them. See: Baseline Document for Thematic Area no. 5, 'Risk Management', Fourth World Water Forum, 16-22 March, 2006. Accessible at www.worldwaterforum4.org.mx/uploads/TBK_DOCS_50_33.pdf. For more statistical studies, see also the United Nations Inter-Agency Secretariat of the International Strategy for Disaster Reduction or the Emergency Disasters Database (http://www.em-dat.net/index.htm).
11. *Impacts of a Warming Arctic*, Arctic Climate Impact Assessment (ACIA), Cambridge Uni-

12. versity Press, 2004 (accessible at http://amap.no/acia).
12. IPCC, *Climate Change 2001: Synthesis Report*, op. cit.; IPCC, *Strategies for Adaptation for Sea Level Rise* (IPCC, 1991); James G. Titus, Vinjay Narayanan, *The Probability of Sea Level Rise* (US Environmental Protection Agency, Washington D.C., 1995).
13. *The European Environment: State and Outlook* (European Environment Agency, 2005).
14. *GEO Year Book 2006* (Division of Early Warning and Assessment, United Nations Environment Programme, 2006).
15. *Impacts of a Warming Arctic*, op. cit. The collapse of the thermohaline circulation is one of the main scenarios that might explain sudden climate change. It is linked to the salinity of the North Atlantic Ocean and the impact of this on the Gulf Stream and the North Atlantic Drift. This option is notably retained in the well-known report of Peter Schwartz and Doug Randall, 'An Abrupt Climate Change Scenario and Its Implications for United States National Security', October 2003. The role and impact of salinity is discussed in H. Hatun et al., 'Influence of the Atlantic Subpolar Gyre on the Thermohaline Circulation', *Science*, vol. 309, pp. 1841-1844, 2005; Ruth Curry and Cecilie Mauritzen, 'Dilution of the Northern North Atlantic Ocean in Recent Decades', *Science*, vol. 308, pp. 1772-1774, 2005.
16. *World Urbanization Prospects: The 2003 Revision* (Department of Economic and Social Affairs, Population Division, United Nations, 2003).
17. Demography is not a problem *per se*, since world agriculture is able to feed about three billion additional people. *World Agriculture: Towards 2015/2030. An FAO Perspective* (Food and Agriculture Organization, 2005).
18. Global impact on developing and developed countries is summarised in *Climate Change Futures: Health, Ecological and Economic Dimensions* (The Center for Health and the Global Environment, Harvard Medical School, 2005).
19. *Poverty and Climate Change: Reducing the Vulnerability of the Poor through Adaptation*, a contribution to the 8th Conference of the Parties to the UN Framework Convention on Climate Change, October 2002; *World Resources, 2005: The Wealth of the Poor - Managing Ecosystems to Fight Poverty*, World Resources Institute (WRI) in collaboration with United Nations Development Programme (United Nations Environment Programme and The World Bank, 2005).
20. i.e., river and lake water and renewable groundwater.
21. *Challenges to International Waters – Regional Assessments in a Global Perspective* (United Nations Environment Programme, Nairobi, Kenya, 2006); Jill Boberg, *Liquid Assets*, op. cit.; *Let it Reign: The New Water Paradigm for Global Food Security* (Swedish International Development Cooperation Agency, 2005).
22. Approximately 60% of global freshwater and 40% of the world's population are located within the 263 international river basins. 60% of these basins are not concerned by any kind of international management framework. *Challenges to International Waters – Regional Assessments in a Global Perspective...*, op. cit.
23. Murl Baker et al, *Conflict Timber: Dimensions of the Problem in Asia and Africa*, Final Report Submitted to the United States Agency for International Development, ARD, 2003; *World Agriculture: Towards 2015/2030. An FAO Perspective*, op.cit.; *Global Forest Resources Assessment* (Food and Agriculture Organization, 2005).
24. A carbon sink is a process that aborbs more CO_2 than it produces. Forests and oceans are natural carbon sinks. Deforestation and burnings can turn forests into net producers of CO_2.
25. *State of the World's Forests 2005* (Food and Agriculture Organization, 2005).
26. *State of the World's Forests 2001* (Food and Agriculture Organization, 2001).

27. Ibid.
28. Christopher Delgado et al., *Livestock to 2020 – The Next Food Revolution* (ILRI, Addis-Ababa, Ethiopia, 1999).
29. 2.4 billion people currently use wood and biomass combustibles for cooking and heating; *World Energy Outlook 2002* (International Energy Agency [IEA], Paris, 2002).
30. Ibid.
31. *World Urbanization Prospects...*, op. cit.
32. *Impact of Europe's Changing Climate*, EEA Report 2/2004 (European Environment Agency, 2004); *Analysis of Post 2012 Climate Policy Scenarios with Limited Participation* (Institute for Prospective Technological Studies, EUR 21758 EN, 2005); *Energy and Environment in the European Union* (European Environment Agency, 2002). The North Atlantic Oscillation could intermittently modify those trends. See James Hurell, Yochanan Kushnir, Geir Ottersen and Martin Visbeck, 'The North Atlantic Oscillation: Climate Significance and Environmental Impact', *Geophysical Monograph* 134, 2003.
33. A 50% reduction in river discharge is expected in those regions by 2070.
34. Southern Europe (France and Austria included) suffered a drop of 30% of crops yields during the 2003 heatwave, see *Impact of Europe's Changing Climate*, op. cit.
35. *Vulnerability and Adaptation to Climate Change in Europe*, Technical Report no. 7/2005 (European Environment Agency, 2005).
36. Heatwaves have become more common in Europe over the past three decades. According to the World Health Organization 'The European climate assessment confirms that Europe has experienced an unprecedented rate of warming in recent decades. From 1976 to 1999, the annual number of warm extremes increased twice as fast as expected based on the corresponding decrease in the number of cold extremes.', *Heat-waves: risks and responses*, Health and Global Environmental Change Series no. 2 (World Health Organization Europe, 2004).
37. *Progress Towards Achieving the Community's Kyoto Target*, Report from the Commission, COM(2005) 655 final, Brussels, 15 December 2005.
38. *A European Strategy for Sustainable, Competitive and Secure Energy. What is at Stake*, Background document, Annex to the Green Paper, {COM(2006) 105 final}, SEC(2006) 317/2, Brussels, 2006.
39. However, due to the rejection of the Kyoto protocol, the level of future US greenhouse gas and aerosol emissions is particularly difficult to predict. See *World Energy Outlook 2005*, op.cit.
40. Lower emissions scenarios predict a rise of 1°C-3°C by 2100 but high emissions rate scenarios forecast a rise ranging between 3.5°C and 7.5°C. In this context, the particular role played by the States in US environmental policy should not be underestimated. Some of them have adopted advanced measures on sustainable development. See Patricio Silva, *Evaluating US States' Climate Change Initiatives*, prepared for the meeting: 'Federalism and US Climate Change Policy, Business and Policy Implications of US States' Climate Actions', CFE-IFRI, Paris, 24 May, 2004.
41. The Rio Grande/Rio Bravo, which separates the US from Mexico, is an area of concern. On the one hand, the US overuse of water already affects downstream discharge of the river. On the other hand, the Las Cruces/El Paso/Juarez region could count 6 million inhabitants in 2025 (currently 2 million). *Challenges to International Waters – Regional Assessments in a Global Perspective ...*, op. cit.
42. See the National Oceanic and Atmospheric Administration website (NOAA – http://www.aoml.noaa.gov/hrd/tcfaq/G4.html).
43. US policy remains in line with the Byrd-Hagel Senate Resolution, (S. RES. 98, 105th

CONGRESS 1st Session), which proscribed US participation in the Kyoto Protocol for the sake of economic competitiveness, and condemned the non-commitment of the developing countries to the process. Nevertheless, federal States tend to adopt, on an individual basis, a stricter approach on environment. See Patricio Silva, *Evaluating US States' Climate Change Initiatives*, op. cit.

44. The concept of gas intensity has been introduced by the US administration: 'Greenhouse gas intensity measures the ratio of greenhouse gas (GHG) emissions to economic output. This new approach focuses on reducing the growth of GHG emissions, while sustaining the economic growth needed to finance investment in new, clean energy technologies. It sets America on a path to slow the growth of greenhouse gas emissions, and – as the science justifies – to stop and then reverse that growth.' *Global Climate Change Policy Book*, The White House, February 2002 (http://www.whitehouse.gov/news/releases/2002/02/climatechange.html).

45. IPCC, *Climate Change 2001: Impacts, Adaptation and Vulnerability*, 2001; IPCC, *Climate Change 2001: Synthesis Report*, op. cit. See also the National Oceanic and Atmospheric Administration website: http://www.noaa.gov/.

46. http://www.whrc.org/southamerica/index.htm.

47. IPCC, *Climate Change 2001: Impacts, Adaptation and Vulnerability*, op. cit.

48. IPCC, *Climate Change 2001: Synthesis Report*, op. cit.

49. Water stress is defined as 1,700 m^3 or less per person annually. Water scarcity implies 1,000 m^3 per person annually.

50. *Global Environmental Outlook 3*, United Nations Environment Programme, 2002.

51. *Challenges to International Waters – Regional Assessments in a Global Perspective...*, op. cit.

52. Sara J. Scherr, *Soil Degradation: A Threat to Developing-Country Food Security by 2020?* (Discussion Paper 27, International Food Policy Research Institute, Washington D.C. 1999); Joachim von Braun et al., *New risks and opportunities for Food Security: Scenario analyses for 2015 and 2050* (Discussion Paper 39, International Food Policy Research Institute, Washington D.C., 2005).

53. Millennium Ecosystem Assessment, *Ecosystems and Human Well-Being: Health Synthesis* (World Health Organization, 2005); *Climate Change Futures: Health, Ecological and Economic Dimensions* (The Center of Health and the Global Environment, Harvard Medical School, November 2005).

54. Asia-Pacific Regional Document', World Water Forum, 16-22 March 2006. Accessible at http://www.worldwaterforum4.org.mx/uploads/TBL_DOCS_107_49.pdf.

55. 97% of Egyptian freshwater resources come from outside and about 70% for Syria.

56. *Middle East and North Africa Regional Water Initiative*, Regional Water Initiative Report (The World Bank Group, Spring 2002).

57. In 2002 Saudi Arabia spent 1.7% of its GDP, roughly US$ 3.4 billion, in this sector whereas at the same time Egypt invested US$ 750 million in water supply and sanitation services. In the case of Egypt, the water sector – including irrigation – represents 20% of the national state budget. Countries of the Gulf Cooperation Council (GCC) have invested as much as about US$ 4.9 billion in water supply services.

58. Irrigated arid land already constitutes a major part (18 out of 26 million hectares) of Near East and North Africa's total irrigated land (FAO, *World Agriculture Towards 2015/2030...*, op. cit). Moreover, food imports already cover more than 50% of the local needs, while 80% of the regional cultivations depend on erratic rainfall.

59. Chirine H. Alameddine, *Le Développement urbain au Moyen-Orient et en Afrique du Nord*, Note sectorielle (The World Bank, August 2005).

60. The notion of 'Green' GDP is increasingly applied in the Chinese debate to subtract from 'Black' GDP the financial implications of environmental degradation, with a difference amounting to more than $100 billion. The World Bank, *World Development Indicators*, quoted in 'L'energia al potere', *Aspenia*, no. 32, 2006. See also Joshua Cooper Ramo, *The Beijing Consensus* (The Foreign Policy Centre, 2004).
61. Under a worst-case scenario, short of implementing major policy programmes to promote efficiency in energy production and consumption, per capita annual emissions levels could reach 1.33 tonnes of carbon by 2020: more than double the per capita emissions of the year 2000 (but still much lower than OECD countries). Dai Yande, Zhu Yuezhong, Jonathan E. Sint, 'China's Energy Demand Scenarios to 2020', *The Sinosphere Journal*, vol. 7, Issue 1, May 2004.
62. According to the Chinese Academy of Sciences (CAS) the total area of China's deserts is 1.57 million square kilometres. Northern arid lands of China are a subject of concern since a large part of them, according to official statements, are entering the first stages of desertification. In Northwest China's Qinghai Province, a rise in temperatures is endangering the glaciers. 17% have disappeared in the last 30 years, causing the loss of 2.39 billion cubic metres of water.
63. Soil erosion, moreover, affects as much as one third of the country.
64. Li ZhiDong, 'Energy and Environmental Problems behind China's High Economic Growth – A Comprehensive Study of Medium- and Long-term Problems, Measures and International Cooperation', IEEJ, March 2003 and Chinese Ministry of Environment, *Report On the State of the Environment In China*, 2004 (http://www.zhb.gov.cn/english/SOE/soechina2004/land.htm).
65. Li Chenggui, Wang Hongchun, 'China's Food Security and International Trade', *China & World Economy*, no. 4, 2002.
66. Yingling Liu, 'Shrinking Arable Lands Jeopardizing China's Food Security', Worldwatch Institute, 18 April 2006.
67. Located in the Beijing and Tianjin municipalities and Hebei, Shandong, Henan and Jiangsu provinces.
68. Nathan Nankivell, 'The National Security Implications of China's Emerging Water Crisis', *China Brief*, vol. 5, Issue 17, The Jamestown Foundation, 2 August 2005.
69. See Norman Myers, 'Environmental refugees: an emergent security issue', 13th Economic Forum, Session III – Environment and Migration, Prague, 23-27 May 2005. For a more contrasted view, see Richard Black, 'Environmental refugees: myth or reality?', Working Paper no. 34, available on the web site of the Journal of Humanitarian Assistance (http://www.jha.ac/articles/u034.pdf).
70. This project aims at linking large water basins to water-consuming regions. Three canals are to be constructed, from the Yangtze river to Western and Northern China. Costs are currently estimated around 61.6 billion dollars. Xinhua News Agency, 5 March 2003.
71. Ti Le-Huu, 'Natural Disasters: overview of recent trends in natural disasters in Asia and the Pacific', *Water Resources* (Journal ST/ESCAP/SER.C/217, December 2005).
72. With many of its cities located along the coast, the Asia-Pacific region remains particularly vulnerable to water-related catastrophes. To illustrate, about 62,000 people were killed in water-related catastrophes between 2001-2005. See: 'Asia-Pacific Regional Document', World Water Forum, 16-22 March 2006. Accessible at: http://www.worldwaterforum4.org.mx/uploads/TBL_DOCS_107_49.pdf.
73. Aerosols could thwart the effects of greenhouse gases, but they will also impact on the

environment. Thus, whereas rises in temperature and precipitation are expected in the long term, locally – notably in India – aerosols are likely to cool temperatures, strengthen thermal inversion and diminish precipitation. The consequent environmental degradation could also jeopardise the predicted increase in India's cereal production. See *The Asian Brown Cloud: Climate and Other Environmental Impacts*, United Nations Environment Programme, Center for Clouds, Chemistry and Climate, UNEP/DEWA/RS.02-3, 2002; G.S. Bhalla et al., *Prospects for India's Cereal Supply and Demand to 2020* (International Food Policy Research Institute, Discussion Paper 29, Washington D.C., November 1999).

74. Mary M. Kent and Sandra Yin, 'Controlling Infectious Diseases', *Population Bulletin*, vol. 61, no. 2, June 2006.

Science and Technology

The field of science and technology (S&T) is evolving rapidly. Its applications and techniques are found in many domains, ranging from health sciences to information technology. Over the past twenty years, S&T discoveries and products have revolutionised the way we live, interact, and work. Looking towards 2025, the trend is likely to continue. At least four key drivers are expected to fuel the process.

> **General Trends**
>
> - Growth in information technology (IT). The centrality of IT in daily life is likely to increase as new IT applications are developed and computing power, data storage, and bandwidth capacity are enhanced.
>
> - Advances in nanotechnology. Nanotechnology, or the ability to control and assemble materials measured in nanometres (a billionth of a metre), may radically transform the way goods are assembled and manufactured.[1]
>
> - Innovations in biotechnology. Advances in biotechnology may improve our capacity to cope with disease, malnutrition, and pollution. Innovations are expected to increase both the quantity and quality of human life in many parts of the world.
>
> - Research and development investments. Current trends in public and private research and development investment point to increased investment in R&D in areas such as nanotechnology, raising the prospect of future breakthroughs.

Synergies among these four enablers will also play a pivotal role. For example, in around 2015, the combination of certain elements of biotechnology, IT, and nanotechnology may yield implantable nano-sized devices that can deliver personalised medicines.[2] By 2025, we may see 'smart' materials that are responsive to physical variables such as light, heat, noise, odours, and electromagnetic fields.[3]

It should be noted that making projections in the field of science and technology is difficult. New discoveries need to be transformed into products and applications before their impact can be appreciated. For instance, when the Internet went mainstream, few could have predicted that it would spawn applications such as instant messaging, online telephony, peer-to-peer sharing systems, and web logs (blogs) further down the line.

Moreover, the amalgamation of different technologies may result in applications that are not immediately evident. The combination of satellite technology and precise atomic clocks, for example, paved the way for positioning, timing, and navigation services. From a different angle, challenges such as high production costs, unexpected societal reactions, legal implications, or other factors may complicate or block product development. With these challenges in mind, what are some of the more likely trends in the field of S&T in the next twenty years?

Growth in information technology

Over the last thirty years, computing power has doubled approximately every two years. Known as Moore's Law, the trend has encouraged the IT revolution and an upsurge in associated technological applications. Studies suggest that Moore's Law will hold for another ten to twenty years before the limits of silicon technology are reached.[4]

In around 2015, industry experts expect chip manufacturers to start moving away from silicon transistors as relevant nanotechnology applications materialise.[5] Prospective replacement materials include carbon nanotubes or silicon nanowires that could provide three times greater performance compared to conventional transistors of the same power level.[6] Closer to 2020, transistors may consist of only a few atoms – introducing the possibility of quantum computing.[7]

The capacity of future computer systems will be staggering. By 2010, Cray hopes to develop a computer system capable of sustaining petaflop performance.[8] This equates to computing speeds that are 1,000 times faster than today's supercomputers that reach teraflop speeds.[9] Around the same time, new data storage techniques such as nanotechnology-enabled memory may vastly increase the capacity to maintain data.[10] With respect to bandwidth, supply is expected to exceed demand in most cases as we approach 2020.[11] The implications are likely to be wide-ranging, affecting how we work and how society is organised. Beyond IT-related impacts, these developments are likely to reinforce the process of globalisation – especially as interconnectivity grows worldwide.

The Internet will also continue to evolve. The large-scale integration of computer systems via high-speed networks will facilitate 'grid-computing' – allowing users to access computing power, storage capacity, data, and software irrespective of their location.[12] The gradual switch towards grid computing may make the personal computer less relevant in the future.[13] Grid computing may also shorten time spent on tasks significantly. For example, with access to centralised computing power, a civil engineer may be able to test an earthquake-resistant bridge design in minutes rather than months.[14]

Advances in nanotechnology

Nanotechnology (NT) is a hybrid science that combines elements of chemistry and engineering. Scientists speculate that major nanotechnology breakthroughs are liable to occur anywhere from five to twenty years in the future depending on specific commercial products and applications.[15] This does not include current uses of nanoscale materials in industries such as electronics, magnetics, cosmetics, and pharmaceuticals.[16] There is an expectation, however, that nanotechnology could be used in nearly half of all new products by 2015.[17]

Nanotechnology is likely to impact on different areas of electronics, medicine, materials, engineering, and the environment.[18] Many believe that NT will be the backbone of the next technology wave as new platforms become smaller and more powerful.[19] If successful, molecular electronics could extend Moore's law well beyond 2015.[20] In medicine, advances in nanotechnology may help improve diagnostic tests, upgrade imaging agents, and enhance cancer therapies. By 2015, semiconductor quantum dot technology based on NT could overhaul chemical labelling, enabling rapid processing for drug discovery, genotyping, and other biological applications.[21] It is estimated that the demand for nanotechnology health care products will exceed $100 billion by 2020.[22]

Breakthroughs in IT and nanotechnology may also have consequences for automation, impacting on processes ranging from commercial logistics to the conduct of warfare.[23] The potential for robotic transport vehicles, for instance, was first demonstrated during the second Darpa Grand Challenge held in October 2004. Using an assortment of sensors, several 'driverless' vehicles were able to navigate a 131-mile desert course.[24] As this technology matures over the next five to fifteen years, an assortment of automated transport and cargo shipments may become widespread.

Innovations in biotechnology

Biotechnology refers to the application of S&T to living organisms in order 'to alter living or non-living materials for the production of knowledge, goods, and services.'[25] While biotech advances have already impacted on certain sectors (e.g. pharmaceuticals), analysts predict that biotechnology will start revolutionising life itself sometime around 2015.[26] It is expected that advances in biotechnology will extend life expectancy in many parts of the world through 'better disease control, custom drugs, gene therapy, age mitigation and reversal, memory drugs, prosthetics, bionic implants, animal transplants' as well as other advances.[27] Overall, effects may be felt in areas such as human health, agriculture, and industry (e.g. biofuels and biorefineries).[28]

By 2025, diagnostic tools may be more predictive, therapeutic interventions more preventive, and healthcare more personalised.[29] The ability to use bioartificial organs is projected to mature between 2010-2025. Stem cell-based therapies for the treatment of chronic diseases such as diabetes, Parkinson's and Alzheimer's may emerge around 2015-2025.[30] Discoveries in biotechnology may also help develop new vaccines against a host of diseases or pathogens.[31]

With respect to agriculture, advances in biotechnology may help usher in a new generation of genetically-engineered crops. Advances to date, focussing on the production of herbicide-tolerant and insect-resistant crops, may gradually give way to crops that are longer lasting and healthier for consumption.[32] For example, some companies are investigating genetic engineering principles to produce foods with lower values of trans fats between 2010 and 2015.[33] Overall, such developments may have implications in areas ranging from poverty reduction efforts (e.g. by developing crops that can grow in formerly unproductive regions) to the promotion of healthier diets to increase life expectancy.[34]

Synergies across the IT and biotech sectors may also enhance existing products such as sensors. In the future, biosensors may be able to probe the environment for specific molecules that are either airborne, solid materials, or liquids resulting in significantly fewer false positives compared to present-day systems.[35] Networking these devices would enable users to monitor different locations on a real-time basis.

Research and development investments

Public and private R&D investments often represent the stepping-stone for future S&T discoveries.[36] Together with other factors, such as the quality and availability of scientific education, they provide important indications of where innovations might take place.[37] While there are no long-term projections, gross expenditure on R&D (GERD) represents a general indicator of R&D investments. In 2002, US contributions towards GERD were 35% of the world total. Asia came in second with 31.5% while Europe came in third with 27.3%.[38] In contrast, some regions captured very small portions of overall GERD. Africa, for example, represented 0.65% of global GERD.[39] It should be noted that certain countries within these regions have increased their GERD investment levels substantially. To illustrate this point: China's GERD/GDP ratio went from 0.6% in 1996 to 1.23% in 2002.[40] This trend is also reflected in the number of patent filings made by individual countries. Between 2000 and 2005, the number of patent applications by Japan, the Republic of Korea, and China rose by 162%, 200% and 212% respectively.[41]

R&D investments in key sectors such as nanotechnology may also shed light on future S&T developments. Overall, government-funded research in nanotechnology increased from $500 million in 1997 to $3.5 billion in 2003 – highlighting the field's growth potential.[42] Currently, the US and Europe are the principal investors in nanotechnology. In the United States, the 2006 National Nanotechnology Initiative (NNI) requested $1.05 billion for nanotechnology R&D from the relevant departments across the federal government – a 137% increase from 2001.[43] Within the EU, €1.3 billion was assigned for nanotechnology research through the 6th Framework Programme (2002-2006).[44] Other countries, however, are taking notice and increasing their efforts in the field. While half of the world publications in nanoscience and nanotechnology come from the US and Europe, countries such as China, Japan, and South Korea account for most of the remaining share.[45]

With respect to human capital formation, a variety of metrics point to US research universities being world leaders in science and technology.[46] Thirty-seven of the top fifty universities (all fields) are located in the United States.[47] Given this prominence, the United States is the principal destination for the world's top foreign students. In 1999, Chinese and Indian-born students accounted for the vast majority, numbering about 33,000 and 23,000 respectively (35% and 25% of the total). Overall, foreign students accounted for almost 50% of all Ph.Ds in engineering, mathematics, and computer science.[48] A more complicated visa process in the aftermath of 9/11 temporarily slowed down the entry pace of foreign students. Presently, increased competition from other countries in Europe and Asia is putting additional pressure on the United States, intensifying the scramble to acquire the best talent.

Challenges posed by S&T developments

It should be acknowledged that there are potential downsides associated with S&T developments – many of which are frequently neglected. Left unattended, their implications could be substantial and long-lasting. In addition, since technical applications tend to outpace existing laws and regulations, there is scope for a range of unintended consequences. Below are examples of challenges associated with certain S&T drivers.

Growth in information technology (IT)

- *Loss of personal privacy* – Improved abilities to maintain large databases and monitor electronic signals may result in substantial loss of privacy. Enhanced sensors and positioning systems may likewise affect privacy negatively.
- *Increased vulnerabilities* – Greater reliance on technology across society may make us more vulnerable to electronic attacks. With more critical infrastructures and systems interconnected, these risks can multiply via so-called 'cascading effects'.
- *Maintenance of the digital divide* – Regional disparities may linger or grow if some regions of the world cannot benefit or tap into the IT revolution. Potential side effects include sustained poverty and/or regional instability.

Advances in nanotechnology

- *Questionable applications* – These may range from the use of NT for harmful purposes to unintended consequences associated with the growing use of automated processes.[49]

Innovations in biotechnology

- *Questionable applications* – These may include issues surrounding human reproductive technologies (e.g. cloning), the use of embryos to procure stem cells, and the use of genetic therapeutics for enhancement purposes.[50] Societal reactions may also surface in other application areas such as the use of genetically-engineered crops.
- *Loss of privacy* – Improved capacities to diagnose (as well as to map genes) may make it easier to gauge personal health histories and predispositions towards disease. Individuals affected may be barred from certain jobs or denied health insurance.
- *Dual-use risks* – Technologies that serve to improve human welfare may also be used to inflict harm. For example, advances in civilian biotechnology may indirectly help certain individuals or groups acquire biological warfare capabilities.

Notes

1. The term sometimes includes the concept of nanoscience.
2. Philip S. Anton, Richard Silberglitt, and James Schneider, *The Global Technology Revolution*, RAND MR-1307 (Santa Monica, California, 2001).
3. Joseph F. Coates, John Mahaffie, and Andy Hines, *2025: Scenarios of US and Global Society Reshaped by Science and Technology* (Oakhill Press, Greensboro, 1998).
4. Some studies predict that silicon technology may last for another ten to fifteen years. See Richard O. Hundley et al., *The Global Course of the Information Revolution: Recurring Themes and Regional Variations*, RAND MR-1680-NIC (Santa Monica, California, 2003), p. xxiii. Others, including Gordon Moore himself, predict that the fundamental limits in the size of a transistor will be reached in around 2015-2025. See Manek Dubash, 'Moore's Law is dead says Gordon Moore', *Techworld*, 13 April 2005.
5. Michael Kanellos, 'Intel sketches out nanotechnology road map', news.com, 25 October 2005. Available at http://news.com.com/ 2102-1006_3-5424766.html?tag=st.util.print.
6. Ibid.
7. The possibility of controlling the spin of electrons for the purpose of enhancing computing power – known as spintronics or spin-based electronics – may materialise soon afterwards. See Michael Kanellos, op. cit.
8. For reference, a petaflop is the equivalent of a thousand trillion floating point operations per second. Information on the Cascade project comes from Cray Fact Sheet, October 2005, available at http:// www.cray.com/downloads/cray_factsheet.pdf.
9. Ibid.
10. Includes Ovonic, Holographic and Nanocrystalline memory markets. See article 'Nanotechnology-Enabled Memory Market to Surpass $7 Billion in 2010, Says New NanoMarkets Report', Nanotechwire.com, 6 February 2006. Accessible at http://nanotechwire.com/ news.asp?nid=2879.
11. Offers of bandwidth are expected to be in the Gigabytes per second (Gbps) range. Advances in WiFi and Ultra-Wideband will also make it possible to upload substantial amounts of information onto portable devices. 'Mirroring' technologies may give the perception of high bandwidth without the need to transport information across networks. See Michael Rader et al., 'Key Factors Driving the Future Information Society in the European Research Area', Technical Report Series, European Commission, Institute for Prospective Technological Studies, Joint Research Centre, September 2004.
12. Ian Foster, 'The Grid: Computing without Bounds', *Scientific American*, vol. 288, no. 4, April 2003. For more on grid computing, see http://www.gridcomputing.com/
13. 'Key Factors Driving the Future Information Society in the European Research Area', op. cit.
14. Ian Foster, op. cit. See also 'Platform 2015: Intel Processor and Platform Evolution for the Next Decade', Intel White Paper, 2005.
15. See website of the National Nanotechnology Initiative, available at http://www.nano.gov/html/res/faqs.html.
16. Among the areas producing the greatest revenues are nanotech applications associated with chemical-mechanical polishing, magnetic recording tapes, sunscreens, automotive catalyst supports, biolabelling, electroconductive coatings, and optical fibres. See National Nanotechnology Institute section on applications and products, accessible at http://www.nano.gov/html/facts/appsprod.html.

17. 'Nanotechnology: Societal Implications – Maximizing Benefits for Humanity', Report of the National Nanotechnology Initiative Workshop, 2-3 December 2003.
18. http://www.nano.gov/html/facts/appsprod.html.
19. Jürgen Altmann, 'Military Uses of Nanotechnology: Perspectives and Concerns', *Security Dialogue*, vol. 35, no. 1, pp. 61-79, March 2004.
20. John Teresko, 'Get Ready for the Age of Nanotechnology', Forbes.com, October 2003, available at http://www.forbes.com/2003/10/02/1002nanotechnologypinnacor_print.html.
21. Philip S. Anton et al, *The Global Technology Revolution*, op. cit.
22. Based on figures presented in 'Nanotechnology in Health Care', The Freedonia Group, May 2005, accessed through *Nanotechnology Now*: http://www.nanotech-now.com/news.cgi?story_id=09445 (30 January 2006).
23. See for example Preston Lerner, 'Robots go to war', *Popular Science*, vol. 268, no. 1, January 2006.
24. See http://www.darpa.mil/grandchallenge/
25. Based on OECD's statistical definition of biotechnology. See http://www.oecd.org/document/42/0,2340,en_2649_37437_1933994_1_1_1_37437,00.html.
26. Philip S. Anton et al, *The Global Technology Revolution*, op. cit.
27. Ibid., p. xii.
28. For example, advances in industrial biotechnology may introduce more cost-effective technologies to convert renewable biomasses such as crops into biofuels. 'Biotechnologies to 2025', Report prepared for the New Zealand Agencies by the Ministry of Research, Science and Technology, January 2005. Accessible at http://www.morst.govt.nz/uploadedfiles/Biotechnology/FutureWatchBookFull.pdf.
29. To illustrate, advances in genomics may help scientists learn how to profile, copy, and manipulate genetic materials of human and plant organisms. See 'Biotechnologies to 2025', op. cit.
30. Ibid.
31. It is worth noting that only two of the thirteen infective Category A agents have vaccines against them. 'Biotechnologies to 2025', op. cit.
32. Andrew Pollack, 'Genetically-engineered Crops: The Next Generation', *The International Herald Tribune*, 15 February 2006.
33. Ibid.
34. Philip S. Anton et al, *The Global Technology Revolution*, op. cit.
35. 'Biotechnologies to 2025', op. cit.
36. Public sector R&D is usually geared towards basic research and the long-term development of new technologies while the private sector focuses on developing and deploying existing technologies in the short- to medium-term. See Brian S. Fisher et al., 'Technological development and economic growth', Abare Research Report, January 2006.
37. This section does not consider current R&D investments that might bear fruit beyond 2025 (such as the prospects for nuclear fusion).
38. *UNESCO Science Report 2005*, 2005. For more information on the EU, see 'Creating an Innovative Europe', Report of the Independent Expert Group on R&D and Innovation appointed following the Hampton Court Summit, January 2006. Available at http://europa.eu.int/invest-in-research/
39. *UNESCO Science Report 2005*, 2005.
40. Ping Zhou and Loet Leydesdorff, 'The Emergence of China as a Leading Nation in Sci-

ence', Forthcoming policy paper available at: www.leydesdorff.net/ChinaScience/ChinaScience.pdf.

41. The top five users of the international patent system were the United States, Japan, Germany, France, and the United Kingdom. The US accounted for 33.6% (a 3.8% increase over 2004) of all applications in 2005. For comparative purposes, Japan, the Republic of Korea and China accounted for 24.1% of all applications in 2005. 'Exceptional Growth from North East Asia in Record Year for International Patent Filings', World Intellectual Property Organization, Press Release 436, Geneva, 3 February 2006. Accessible at: http://www.wipo.int/edocs/prdocs/en/2006/wipo_pr_2006_436.html?printable=true.

42. Michael Kanellos, op. cit.

43. 'The National Nanotechnology Initiative: Research and Development Leading to a Revolution in Technology and Industry', Supplement to the President's FY 2006 Budget, March 2005.

44. This does not include funding provided at the EU member state level. 'Key Factors Driving the Future Information Society in the European Research Area', op. cit.

45. Ping Zhou and Loet Leydesdorff, op. cit. See also Carsten A. Holz, *China's Economic Growth 1978-2025: What We Know Today about China's Growth Tomorrow*, Centre on China's Transnational Relations, Working Paper No. 8, The Hong Kong University of Science and Technology, July 2005.

46. *UNESCO Science Report 2005*, 2005.

47. *Academic Rankings of World Universities 2005*, published by Institute of Higher Education, Shanghai Jiao Tong University. Accessible at http://ed.sjtu.edu.cn/rank/2005/ARWU2005TOP500list.htm.

48. *UNESCO Science Report 2005*, 2005.

49. An international dialogue has started to discuss the best ways to ensure a responsible development of nanotechnology. For a copy of the proceedings of the first workshop (2003) see http://www.nano.gov/nni_societal_implications.pdf.

50. Philip S. Anton et al, *The Global Technology Revolution*, op. cit.; 'Biotechnologies to 2025', op. cit.

Part II
The Regions

Eurasia and Russia

Twenty years from now, the former Soviet Union as a concept will no longer be meaningful. Over time, this area of fifteen states will have shattered as a unified region into several new regions: a new Eastern European region on the EU's borders; the South Caucasus, heavily tied to the Black Sea region; and the Central Asian region, associated closely with China and South Asia. The Russian Federation will remain as a pivot linking these various regions together.

Eurasian trends

A diverse region

Regional differences reside at three levels which, two decades from now, will be more pronounced. First, the region will show sharp variations in political and economic structures, ranging from democratic polities and market economies in Eastern Europe (Ukraine, Belarus, Moldova and parts of the South Caucasus) to more authoritarian state-regulated societies and economies in Central Asia. The Russian Federation is likely to have developed a state-led market economy within a democratic political system that has strong illiberal features. Second, by 2025, the states of Eurasia will show differences in levels of good governance and stability – independently of their regime type (democratic or semi-authoritarian). Levels of healthy governance will range from strong performance from parts of the European former Soviet Union to weak and unhealthy governance in Central Asia.

Finally, in terms of foreign policy orientation, the region will become heavily diversified. The states in the European region, including also Georgia and possibly Armenia and Azerbaijan, will become more closely tied to the EU zone and to NATO. The Central Asian states will be more oriented in the Asian direction, towards China especially in trade relations. Russia is likely to have developed a more varied foreign policy after its strong Western focus in the 1990s, with deep trade and political ties in the Asian arena in addition to economic and political links with Europe.

The political evolution of Eurasian states

At the same time, despite such diversity, the states of Eurasia will face similar questions at four levels. The first question will concern the continuing process of state building across the region. In 2006, the final status of the post-Soviet states – their borders, their constitutional structures as federations, confederations or unitary states – remains incomplete. In the cases of Georgia, Moldova, Azerbaijan and Russia, separatist and territorial conflicts remain unresolved, which pose fundamental challenges to the final status of their states. By 2025, it is likely that some of these conflicts will have been resolved (Moldova, Azerbaijan and Russia). However, new challenges will arise, most likely in Central Asia regarding border questions between states and the balance within these states between regional and central power. Overall, the states of Eurasia are likely to range from *stronger* states (such as Russia and the European states) to *brittle* states (in the South Caucasus and Kazakhstan) to *weak* states (in Central Asia). The possibility of the region containing *failed* states, such as Tajikistan, cannot be ruled out.

A second question will concern the political regimes of the states in the region. The premise underlying international approaches to the fifteen post-Soviet states in the 1990s viewed these countries as undertaking a transition process towards democratisation and a market economy. By 2006, the emergence of varied political regimes across the region – ranging from strong authoritarian regimes in Belarus and Turkmenistan to fledgling democracies in Ukraine and Georgia – has undermined this overarching premise. Twenty years down the line, Eurasia's political regimes will be ever more diverse. In some cases, such as Belarus, authoritarian regimes are likely to have fallen. The Central Asian states, however, may go through difficult periods of succession following the departure of their current leaders, which may cause serious tensions, perhaps even social unrest. By 2025, stronger democratic polities are likely to have emerged in Ukraine and Moldova, as well as Georgia and Armenia, with more developed multi-party systems and better defined constitutional structures. However, these states, including the Russian Federation, are likely also to demonstrate strong illiberal tendencies, in terms of media control and levels of corruption.

Economic challenges

Third, the Eurasian states will face difficult economic questions by 2025. The changes that have occurred in the economic systems of these countries since 1992 are radical in terms of privatisation and structural economic reform.[1] By 2025, however, the economic development of the Eurasian states will face three inter-related challenges, which are already currently at play. The first concerns whether and how these countries will have offset their resource dependency, either internally, in terms of economic diversification away from energy-driven growth, or externally, in terms of addressing their reliance on external sources of energy supply and developing effective service industries. Indeed, some of these countries, such as Russia, Kazakhstan, Azerbaijan, and Turkmenistan, will face the challenge of building healthy economies that are not solely dependent on the development and export of their natural gas and oil reserves. For other countries, such as Armenia, Georgia, Belarus, Ukraine and Moldova, the challenge will be to diversify their reliance on external energy sources away from their present dependence mostly on Russia. Energy resources are likely to continue to both unite and divide the region – uniting it in terms of the transportation of energy to local and world markets, but dividing it also between energy producers and dependent users.

Another economic challenge will centre on the investment climate in the Eurasian economies, in terms of the rule of law, protection for foreign investment, and levels of corruption. In 2004, for example, Russia's FDI was $9.4 billion, but the net capital outflow was $7.8 billion, leading the EBRD to state that 'the accumulated stock of FDI amounts to about 6.5% of GDP, only a fifth of the average level of the other European transition economies.'[2] Without serous improvements in the investment climate, foreign direct investment in 2025 will most likely remain low, and sectorially targeted.[3] A final economic challenge will be linked to the role of the state in the economies of the region. In 2006, the trend points towards an increasing role for the state in local economies, particularly in declared 'strategic sectors' such as energy. A strengthening of the role of the state, especially if the state is riddled with corruption, may have deleterious consequences, as it will increase the opportunities for rent seeking, limit foreign investment and have negative effects on overall economic performance.

Demographic trends

The fourth question facing the Eurasian states in 2025 is likely to centre on developments in their societies, and in state-society relations. Demographic trends will produce pressures on all of the states for different reasons and with different results. The Russian Federation is facing a steep demographic decline (conservative estimates point to a loss of one million people a year through to 2020[4]) due to low fertility and high mortality rates. By contrast, the Central Asian states will face increasing difficulties stemming from higher population growth rates, and especially the difficulties associated with a 'youth bulge.'[5] In the South Caucasus, as well as the other Eurasian states (Tajikistan especially), demographic difficulties will stem from the impact of continuing out-migration for economic reasons.

Sources of instability

Overall, by 2025, Eurasia will feature a number of difficulties, articulated around the following points:

Relationship between states and societies

The relationship between state structures and societies by 2025 will vary from relatively healthy relations in the European part of the region and some countries of the South Caucasus, to heavier control in the Central Asian states. Societal pressures for greater political participation will take different forms across the region, from inclusive challenges in the European region to more turbulent pressures in Central Asia, as in Russia perhaps also. State structures will have different capacities and will react differently to potential threats that arise at societal levels. In such circumstances, tensions may arise in state-society relations, especially as revisionist ideologies may come to challenge existing regimes. Across Eurasia, such ideologies are likely to have two faces in 2025: an increasingly intolerant nationalism and the rise of political Islam in parts of Central Asia and the Russian Federation.[6] The danger is that the heavy-handed repression of Islamic extremist movements in Central Asia is likely to produce a backlash. Both of these movements may develop in ways that challenge the *status quo* of the early post-Soviet period and, perhaps even, overall state authority.[7]

Relations between haves and have-nots

By 2025, the states in Eurasia will become divided according to levels of economic prosperity, demographic trends, educational and health opportunities and overall relationship to globalisation. According to a 2005 World Bank Report, poverty levels have declined across the region since 1998, driven by rising economic growth rates.[8] However, poverty will remain a problem in many Eurasian states in 2025, including in middle-income countries such as Russia, Belarus and Kazakhstan. This poverty will be reflected in increasing regional disparities inside these countries between living standards and job opportunities in the capital cities and those available in the regions.[9] In so far as differences between countries are concerned, a first group, including some of those states that will develop close associations with the EU zone, may succeed in developing more stable economies and governance structures that can provide educational and health resources for their societies (Moldova, Ukraine, Russia, Kazakhstan, Georgia). Another group lies at the other end of the spectrum, consisting of deeply weak and impoverished states that have worrying demographic trends and provide little in terms of educational and health resources (Kyrgyzstan, Tajikistan, regions *within* other states).[10] Such weak states may also pose challenges to regional security as the sources or transit points for international criminal networks dealing in the smuggling of illicit goods, especially narcotics.[11]

Enduring and new zones of tension

In addition to the possibility of still unresolved conflicts in the region (most probably in Georgia), an analysis of current trends shows that new zones of tension will become clear by 2025. These areas will feature a volatile mixture of entrenched economic dislocation, widespread poverty, poor central state management and high social mobilisation – to produce deep challenges to some states and sub-regions of Eurasia. One should note in particular the worrying trends at all of these levels in Russia's North Caucasus, rising tensions in Central Asia's Ferghana Valley and the possibility of collapsed state control in parts of Tajikistan.

Russian Federation trends

After a decade of severe economic decline, the Russian Federation in 2006 has never looked more strong or confident. Following the 1998 collapse of the Russian rouble, the economy effected an exceptional turnaround. From a high point of 7.3 percent GDP growth in 2003, the World Bank estimated that the rate remained strong at 6.4 percent in 2005.[12] Russia's GDP stood in 2005 at US$ 581 billion. The government has pursued a stable macro-economic policy, an internationally lauded fiscal policy, and launched structural reforms in vital sectors of the economy. Industrial productivity has also increased relative to the decline of the 1990s. Energy production and exports have driven much of the growth. The Russian Federation holds roughly 6% of the world's oil reserves, mainly in Western Siberia.[13] Although estimates are particularly controversial, its oil production is projected to increase by 21% to 11.1 million b/d in 2030. Russia is also the world's largest gas exporter, with estimated reserves of 47.8 billion m³ and current production estimated at around 616.5 billion m³.[14] In addition, relative to the political volatility of the 1990s, Russia's constitutional and federal structure had stabilised by 2006, with the rise of a stronger executive and more potent central power.

Optimistic projections for Russia's development have placed the country in the so-called BRIC grouping, composed of Brazil, Russia, India and China.[15] In this analysis, by 2050 the BRIC economies could be larger than the G6 in dollar terms, with the Russian GDP projected to increase to $US 2,264 billion by 2025 (larger than France and Italy).[16] Such a best-case scenario is dependent on Russia building the conditions for growth through sound macroeconomic policies, strong and stable political institutions, openness to trade and foreign direct investment as well as high levels of education. None of these steps can be taken for granted.

Domestic developments

Until 2025, Russia's development will be conditioned by a number of uncertainties. These uncertainties reside at the political, economic, constitutional and social levels.

An 'illiberal' democracy in the making?

Uncertainties surround the future trajectory of the Russian political system. After 1999, Vladimir Putin succeeded in consolidating a deeply enfeebled state by strengthening the so-called 'power vertical' of executive central power. In so doing, the Russian president declared that Russia would follow the path of crafting a 'sovereign democracy' that is fitting for Russian conditions. For many external observers, Russia's 'sovereign democracy' has raised concerns. To all intents and purposes, the Russian political system has a democratic form, with regular elections, ensured constitutional freedoms as well as checks and balances, but the practice is quite different.[17] Russia's 'sovereign democracy' has translated into weakening political pluralism, increased state control over the media sector, the arbitrary application of the rule of law, and pressures to control independent civil society. In 2006, the Russian political system featured strong elements of what has been called 'illiberal democracy' – combining a liberal democratic form with illiberal content.[18] Certainly, after the years of volatile politics under the rule of Boris Yeltsin, greater centralisation under Vladimir Putin provided vital stability to the Russian political system. However, the longer-term impact of a 'managed democracy' is far more uncertain. By 2025, the dislocation between a centralising state and an increasingly alienated society could produce the rise of revisionist political figures with ideologies that challenge the *status quo*. In particular, the danger in Russia is of the rise of intolerant and exclusive nationalism.

Reasserting centralism

At the constitutional level, the Russian Federation is likely to face continuing questions over the next two decades related to the structure and distribution of power across the country. According to the Constitution, the Russian Federation is composed of 89 subjects, including 49 'regions' and 21 'autonomous republics' (which are ethnically defined). Vladimir Putin reversed the policies of Boris Yeltsin, which had devolved powers from the centre to the federation subjects, through greater control over elections at the regional level and strengthening the power of the executive centre. In many cases, reinforced central power has worked to the benefit of stability, but not everywhere.

The North Caucasus has emerged as a fracture point in federal relations, where deep poverty, high relative population growth, poor central policy and increasing social mobilisation paint a deeply worrying picture. The challenges are different but no less serious in Russia's Far East. For the foreseeable future, therefore, uncertainty is likely to remain in the balance of centre-regional power in Russia's federal relations, with political tensions remaining high in the North Caucasus. It is not unthinkable that some regions and republics may seek greater autonomy in their affairs, especially if they have natural resources of value to world markets (as is the case in parts of Siberia). In other cases, centre-region relations may face tensions because of deepening – whether voluntary or not – regional ties with neighbouring countries.[19]

Economic challenges

Russia's economic development also faces deep uncertainty. The rising importance of energy production and its export for Russia's economic growth raises several challenges for the Russian government. It has been estimated that Russia's oil and gas sector may have accounted for up to 25 percent of GDP in 2003, while employing less than 1 percent of the population.[20] First, future economic growth will be affected by the lack of serious diversification away from resource-dependent development. In this, the Russian leadership would need to undertake wide structural reforms that would open the path to investment in and development of non-energy related sectors, such as industrial goods.[21] The management of Russia's Stabilisation Fund, based on energy windfall profits, will be of crucial importance for such diversification.[22] Second, the current trend of increasing public control over the energy sector may have deleterious long term effects, in terms of increased corruption and collusion, poor management, increased politicisation and overall decreasing efficiency.

Third, the Russian energy sector requires substantial reform, which may be more difficult to achieve in conditions of high state control. Not least, it is estimated that the energy sector as a whole (including exploration, development and maintenance of nuclear, coal, heat, electricity, gas and oil) will require some €715,000 million between 2003-2020.[23] Such massive investment will depend heavily on the effective openness of the overall Russian economy, the development of firm and indiscriminate rule of law and clearly defined property rights.[24] Developing an effective banking sector will also be vitally important for developing internal investment.[25] Attenuating corruption is another strategic challenge, affecting all sectors of the economy and their development.[26]

Social challenges

Finally, the Russian Federation will face a range of social challenges by 2025. First, Russia faces a profound demographic challenge. Despite a slight migration inflow and a marginal rise in the fertility rate (1.40 to 1.58), over the coming twenty years Russia's population is likely to fall, from 143.2 million to 129.2 million people by 2025. The share of working age population will probably remain quite high (around 60%); but the very low increase in the young-age category (around 16% of the population for the 50 years to come) and consequently, the net increase of the share of elderly people (from currently 17.1 to 24.3% by 2025) could create serious social problems. At the same time, the country has one of the highest rates of HIV infection in the world, which may have wide social and economic consequences if current rates are maintained. In addition, Russia will face the problem of distributing the impact of economic growth in terms of living standards and employment prospects across the country's varied regions and republics. Without deep structural changes, Russia may present an extremely diverse social-economic picture, with some cities and regions featuring global 'first world' characteristics while others present attributes closer to the 'third world.'

Russia and the world

Russia's strategic outlook

Enduring economic weakness and internal challenges combined with residual and new areas of economic and political strength will most likely lead Russia to adopt a broadly *status quo* orientated foreign policy. For much of the 1990s, in the Russian view, Russia had been subjected to the forces of globalisation with little control over them. As much as possible, Russia will seek to exploit residual elements of its previous superpower status and new areas of strength to ensure that the country becomes more an 'actor' of globalisation and less its 'victim.' Within the overall objective of retaining a great power status, the two main objectives of Russian foreign policy are likely to be, first, that of creating an external environment that is predictable and favourable to Russia's internal transformation, and secondly, to strengthen Russia's sovereignty and freedom of manoeuvre in international affairs. A number of policy lines are likely to stem from these overarching objectives.

Playing a global role

First, the Russian government will continue to insist on maintaining a strong 'voice' in as many international organisations as possible. The UN Security Council has pride of place in this vision, as the foundation pillar of international regimes and international law, and as the only forum ensuring Russian influence on all major international questions. In addition, Russia will seek membership of the World Trade Organization as well as closer association with key regional organisations. In Europe, NATO and the EU, despite current difficulties, will continue to receive significant Russian attention at the political and economic levels. In Russian strategic thinking, the European orientation is vital because of the importance of economic ties and also because Europe is seen as source of security solutions rather than problems for Russia – in contrast to the perceived threats arising from the southern direction. Russia's European focus is likely to be balanced, however, by Russia's desire to become associated with other regional organisations, either as a leader, such as with the Shanghai Cooperation Organisation, or a member/observer, such as with ASEAN.[27]

Second, Russian foreign policy is likely to continue its movement away from the predominant Western and European focus pursued in the 1990s to adopt a more global posture. Europe will remain a key interlocutor for Russia, not least because of the scale of economic exchanges with European partners and because of the importance of Europe as a leader of globalisation. However, Russian foreign policy is likely to become increasingly diversified towards new states that arise as regional and global leaders by 2025, notably towards China and India at the global level and towards Turkey at the regional level. Russian foreign policy is likely to continue to be multi-vectored with the aim of building a multipolar world. The United States will retain its central position in Russia's vision of world affairs, as the single greatest world power, and Moscow will seek to maintain elements of a special relationship with Washington.[28] However, the high point of US influence over Russian internal and foreign policy of the 1990s will not be repeated.

Russia's assets

Third, Russian foreign policy will seek to make maximal use of residual and new elements of strength. The current emphasis on energy in Russian foreign policy as a factor of influence is likely to become reinforced, with Russia seeking to become a truly global energy player through diversified relations away from Europe towards Asia and the United States. Russian strategy will be to develop a more balanced and competitive energy profile, in terms of the production of energy (LNG will increase in importance) and in terms of the orientation of its exports (Europe will remain central but Asia and the United States may also become important markets). In addition, Russia seeks to develop modernised armed forces, in parallel with a competitive military industry. By 2025, Russia aims to have modernised its armed forces in two directions: first, through the renewal and modernisation of its strategic deterrence, and second, through the development of professional and mobile forces able to protect Russian interests inside the country and on its borders. The use of force is likely to retain a role in Russian foreign policy, especially in the former Soviet Union. In addition, the Russian government seeks to retain a leading position in world arms' sales.[29]

The post-Soviet space

Finally, Russian foreign policy will still attribute special importance to the former Soviet Union. The region is important for the Russian economy as a source of labour, a market for Russian products and an enduring (if decreasing) source of energy transferred through the Russian pipeline network. On the whole, Russia's presence in the region will become more diversified, ranging from a regionally-based forward defence policy in Central Asia to strong bilateral ties with Armenia and Azerbaijan, as well as with Belarus and Ukraine. Relations with Moldova and Georgia are likely to remain difficult, with Russia still conducting coercive policies towards the two states. In general, Russian policy will likely become economically-driven throughout the region, as Russian business increases its presence as well as through the development of regional and sub-regional free trade areas.

In 2025, therefore, Russia stands to be a more global player than it was after the collapse of the Soviet Union. Increasing room for manoeuvre will also make Russia a more sovereign actor on the international stage, more able and willing to defend its interests and ensure that its 'voice' is heard. In the former Soviet Union, Russian foreign policy will most likely have a post-imperial direction seeking hegemonic influence but not territorial control. Such a posture may continue to include coercive policies if Russian interests are seen to be threatened.

In conclusion, the central question driving Russia's future will concern the tension between the country's trend towards an increasingly authoritarian centralism and its openness to globalisation. This tension will be played out at several levels that are closely linked. At the federal level, this tension will consist of balancing central control from Moscow and its federal institutions with the powers of Russia's numerous republics and regions. At the political level, Russia's future will be driven by tensions between greater political control, even of an authoritarian nature, by the state and the emergence of a looser and more democratic form of polity. At the economic level, this driving question translates into a tension between an economy that is becoming reliant on the production and export of energy, itself increasingly controlled by the state, and the diversification of the country's economic structure. In managing these multiple tensions, it is possible that Russia may feature elements of an Asian model of political and economic development, including a mix of contrasting aspects. On the one hand, strong centralised political control, including state control of strategic sectors of the economy. On the other, tentative economic diversification and a limited openess to those dimensions of globalisation that are considered to be in Russia's interests. Balancing these tensions will not be an easy task.

Notes

1. *The Development Challenge – Europe and Eurasia*, USAID: http://www.usaid.gov/policy/budget/cbj2004/europe_eurasia/.
2. For an excellent study of FDI in Russia, see *Russia: Investment Destination* (Foreign Investment Advisory Council, March 2005): http://www.pbnco.com/fiacsurvey/.
3. Barriers to FDI are seen to include, most importantly, corruption, a weak legislative and enforcement regime, administrative barriers, inadequate legislation and the selective application of laws; see ibid.
4. *Eurasia 2020. Global Trends. 2020 Regional Report* (Paper associated with the NIC 2020 Project) April 2004: www.dni.gov/nic/PDF_GIF_2020_Support/ 2004_04_25_papers/eurasia_summary.pdf.
5. On the 'youth bulge', see 'Youth in Central Asia: Losing the New Generation,' International Crisis Group, *Asia Report* no. 66, October 2003. For example, Uzbekistan has registered an average population growth of 3% per year: see http://www.umid.uz/Main/Uzbekistan/Population/population.html.
6. The role of Islam, culturally and religiously, has increased since the Soviet collapse across Central Asia and in parts of Russia. This rise has been accompanied also by the development of radical political Islamic groups, which have challenged state authority at various points since 1992 (for example, during the Tajik civil war, 1992-1997, in Uzbekistan, 1999, 2000, and in the Russian North Caucasus during the first and second Chechen wars).
7. On this question for Central Asia, see 'Is Radical Islam inevitable in Central Asia? Priorities for Engagement,' International Crisis Group, *Asia Report* no. 72, December 2003.
8. See *Growth, Poverty and Inequality in Eastern Europe and the Former Soviet Union* (World Bank Report, Team Leader: Asad Alam, October 2005), available: http://web.worldbank.org/wbsite/external/countries/ecaext/0,,contentMDK:20627214~pagePK:146736~piPK:146830~theSitePK:258599,00.html.
9. Of particular note in Central Asia in this respect is the region of the Ferghana Valley.
10. UNDP Human Development Reports for 2005 place Tajikistan as the worst performer in 2005 in Eastern Europe and the CIS, placed at 122 out of 177 countries, followed by Kyrgyzstan ranked at 109 of 177, Moldova at 115 of 177, Georgia at 100 of 177, Ukraine at 78 of 177 and Russia at 62 of 177. See *Human Development Reports 2005*, http://hdr.undp.org/statistics/data/countries.cfm.
11. For an overall picture, see Mark Galeotti (ed.), *Russian and Post-Soviet Organized Crime* (Ashgate, 2002), and for Central Asia in particular, see Erica Marat, 'Impact of Drug Trade and Organized Crime on State Functioning in Kyrgyzstan and Tajikistan,' *China and Eurasia Forum Quarterly*, vol. 4, no. 2, 2006.
12. *Russian Economic Report – April 2006* (World Bank, Moscow Office, Economics Unit).
13. Jan Leijonhielm and Robert L. Larsson, *Russia's Strategic Commodities: Energy and Metals as Security Levers*, Swedish Defence Research Agency, November 2004, p. 14.
14. http://eng.gazpromquestions.ru/page7.shtml. See also: Marina Kim, 'Russian Oil and Gas. Impacts on Global Supplies to 2020', *AustralianCommodities*, vol. 12, no. 2, June Quarter 2005 and e.g., IEA, *World Energy Outlook 2005*, op. cit.
15. See Dominic Wilson and Roopa Purushothaman, *Dreaming with BRICs: The Path to 2050*, Goldman Sachs Global Economics Paper no. 99, 1 October 2003.
16. For all of its overly optimistic assumptions, the BRIC scenario has been accepted as the basis of analysis by the European Round Table of Industrialists; see *Seizing the Opportu-*

nity: Taking the EU-Russia Relationship to the Next Level (ERT, May 2006).

17. *Russia's Wrong Direction: What the United States Can and Should Do* (Council on Foreign Relations, March 2006).
18. Fareed Zakaria, 'The Rise of Illiberal Democracy,' *Foreign Affairs* vol. 76, no. 6, November/December 1997.
19. Most notable is the case of the region of Primorskii Krai, situated on the border with China, with which economic ties have deepened, as have the number of visits by Chinese nationals for economic reasons to the Russian Federation. For early trends in this area, see Mikhail A. Alexseev, *Instrumental Internationalisation: Regional Foreign and Security Policy Interests in Primorskii Krai* (Swiss Federal Institute of Technology, Working Paper no. 18, March 2002, Zurich).
20. See Russia *Country Analysis* (Energy Information Administration, US Government: http://www.eia.doe.gov/cabs/Russia/Background.html)
21. Andrew Dean, *Challenges for the Russian Economy* (OECD Economics Department, Moscow, 7 July 2004).
22. In 2006, the Stabilisation Fund stood at $62 billion. See 'The Russian bear is back and this time it's gas-powered', *The Guardian*, 13 May 2006, as well as the *BOFIT Russia Reviews* (Bank of Finland; http://www.bof.fi/).
23. Christian Cleuntinx, *The EU-Russia Energy Dialogue* (DG for Energy and Transport, European Commission, Vienna, December 2003).
24. Under Vladimir Putin, Russia's attention to intellectual property rights has improved dramatically. However, increasing state control over elements of the energy sector has weakened foreign property rights in this vital sector. On 13 June 2006, the Russian Natural Resources Minister stated that Russia planned to further limit participation by foreign groups in developing oil and gas fields. Such limits may pose obstacles to the investment for modernisation that the Russian energy sector will require over the period to 2025. See *Radio Free Europe/Radio Liberty Newsline* (vol. 10, no. 108, 14 June 2006).
25. On this question, see Stephan Barisitz, *The Evolution of the Russian Banking Sector since Perestroika* (Central Bank, Republic of Austria, 2004: http://www.oenb.at/en/geldp_volksw/zentral_osteuropa/banksecfmstab/barisitz_1_2004.jsp).
26. Corruption is seen as the greatest obstacle to foreign direct investment in Russia; see *Russia: Investment Destination* (Foreign Investment Advisory Council, March 2005): http://www.pbnco.com/fiacsurvey/.
27. Russia is indeed a Dialogue Partner with ASEAN as well as an observer in the Organisation of Islamic Conference.
28. A burgeoning energy relationship between Russia and the US is likely to cement the relationship further, even if projected scales of energy Russian exports towards the US will remain lower than those towards Europe.
29. In 2005, Russian arms sales amounted to $6 billion, according to the Russian Deputy Prime Minister and Defence Minister, Sergei Ivanov; cited in *Moscow News*, 22 March 2006: http://www.mosnews.com/news/2006/03/22/armsorder.shtml.

The Middle East and North Africa

The Middle East and North Africa region consists of 18 diverse Muslim countries from Morocco to Iran. It can be divided into two dissimilar halves, the Maghreb (i.e. North Africa), west of Egypt, and the Mashreq (the Middle East or West Asia), east of Egypt. A belt of adjacent countries is connected to a greater or lesser degree with either or both of these sub-regions. The most important of these countries are Sudan and Somalia, Afghanistan and Pakistan and, increasingly, Turkey. Structural and political developments in the Middle East and North Africa pave the way for growing instability. On the one hand, stability in this region is crucial to the security and prosperity not only of neighbouring areas, including Europe, the Caucasus, Central Asia and the Indian sub-regional complex, but also of the world at large. On the other hand, the region harbours unparalleled potential for crisis, including steady demographic expansion, worsening environmental conditions, stagnating economies and endemic political, ethnic and religious tensions whose resolution is not in sight. The intertwined nature of the ideological (i.e. fundamentalism) and structural (i.e. energy, demography, environment) drivers of potential conflict is another distinctive feature of this region. Although, although the urgent requirement for socio-economic reform, good governance, and regional frameworks of cooperative security is apparent, little progress is being made in that direction.

Regional issues

Energy

Home to the largest oil and gas reserves in the world, the Middle East currently produces 28% of the world's oil. This share should rise to 43% in 2030. By 2030, oil production from the entire MENA region is expected to increase by 74% from 29 mb/d to 50.5 mb/d, and gas production to triple and reach 900 bcm in 2030.[1] The countries of the region (and most particularly Saudi Arabia, Iran and Iraq) are expected to account for around 50% of the EU's oil needs (they currently provide 45%). European gas dependency is, for the time being, less pronounced. The potential increase in European imports from the region may depend on the energy policy of Russia and on the development of transport infrastructures (LNG/pipelines). Since this is the only region with a strong excess oil production capacity, exports from MENA, and notably the Gulf countries, will be vital to sustain the rapid economic growth of other emerging powers. Around 50% of both China's and India's oil demand is supplied by the Middle East, while the flow of gas from Iran and other Gulf countries to China and India might grow via a much-discussed overland pipeline across Pakistan and/or in liquefied form, through the Indian Ocean. Alternative routes from Iran through Central Asia are also being explored.

The challenge of governance

Driven by the rise in oil prices, GDP growth in the region has climbed to 5.5% in 2004 and almost 5.7% in 2005,[2] twice as much as the average rate of growth in the 1990s.[3] From an economic standpoint, accruing oil and gas revenues represent a window of opportunity for the region. Sound governance and structural reforms will, however, be essential to reap the benefits of oil and gas wealth, promote the diversification of the economy and sustain growth. On the whole, whether rich in hydrocarbons or not, all MENA states are confronted with fundamental governance challenges. These include improving the quality of public administration, the rule of law and the accountability of the public sector to citizens and private economic operators.[4] Structural reforms are urgently needed to open up to trade and investment, develop the private sector,[5] and diversify the economy (and thereby limit dependence on oil and energy prices).

Governance reforms are an enabling condition for structural economic reforms. The latter may be painful in the short term, and will require strong political will. Political tensions in most of the region, however, will hardly encourage local regimes to introduce much-needed reforms (such as those of the labour market and banking system, or the further privatisation of public services) which could weaken their domestic power base. As a consequence, the business environment of most MENA countries is likely to remain relatively unattractive. Foreign direct investment, which in 2005 amounted to $9.1 billion and accounts for 0.9% of global FDI (3.8% of FDI to developing countries),[6] is mostly directed to the energy sector in resource-rich countries.

While these governance shortcomings are likely to affect the whole region, however, an important distinction should be drawn between resource-rich and resource-poor countries. Rising prices have resulted in additional oil receipts for the MENA oil-exporting countries worth a total of 39% of GDP over the period 2002-2005.⁷ MENA countries have reacted differently in profiting from this windfall. Large oil- and gas-rich countries, such as Algeria and Iran, have developed a rent economy that stifles further reform and diversification, with high unemployment, a predominant public sector and protective trade barriers. Resource-rich countries around the Gulf, with a much smaller population, have pushed ahead with further privatisation and have greatly improved their infrastructures. For resource-poor countries, on the other hand, sound economic policies and governance reforms are even more pressing, as their growth will depend on better integration in international trade and investment patterns.⁸

Yet another distinction, whose implications go beyond the purely economic dimension, should be pointed out: that between labour-abundant and labour-importing countries. The latter are concentrated in the Gulf region, whose migrant population represents the large majority of the workforce and, in some cases (e.g. UAE and Qatar), of the population itself. The Gulf countries have decided to severely restrain further immigration and to downsize the pool of foreign workers,⁹ a move which is likely to entail very serious social costs and affect the countries benefiting from remittances from the Gulf. Broadly speaking, labour-abundant countries will face the opposite challenge, namely the need to create enough employment opportunities for an exploding population.¹⁰ Demographic expansion is likely to pose the biggest structural challenge for the stability of the region as a whole over the next two decades.

Demographic and environmental trends

By 2025, although fertility rates will progressively decline, the population of MENA will grow by 38%, from 388 to 537 million, with an increase in the working age population of 40% in North Africa and as much as 50% in the Middle East. The working age population will expand by 80 million people (for a total of 185 million people) by 2020. According to the World Bank, roughly 100 million additional jobs would need to be created in the next twenty years to absorb the workforce.¹¹ It is estimated that an economic growth rate of 6 to 7%, maintained over many years, would be required to generate sufficient employment opportunities. Should oil prices fall, and the sluggish performance of the non-energy sectors of the economy continue, unemployment is expected to affect 50 million people (27% of the envisaged overall workforce) within two decades.¹² Things will not be made easier by the growing market-shares acquired by Asian countries, including China but also India, Indonesia and Bangladesh, at the labour-intensive end of the manufacturing industry.¹³ On balance, with the possible exception of some small Gulf countries such as Bahrain and UAE, specialising in financial services, port facilities and tourism, MENA countries risk remaining marginal to economic globalisation, if energy exports are not taken into account.[14] The biggest financial flow directed to countries in the region consists of remittances from their expatriate nationals in Europe or in the Gulf, which amounted to $20.3 billion in 2004.

The deterioration of environmental conditions will represent another challenge for the region. 87% of the MENA surface consists of desert, and that share is bound to expand due to global warming, which induces higher temperatures and lowers precipitations. Despite considerable public investment in potable water supply services,[15] deteriorating environmental conditions, compounded by sustained demographic growth, will lead to a serious water shortage problem. The current annual per capita water availability of about 1,200 m³ (compared to a world level of 7,000 m³) may well drop to 550 m³ by 2050.[16] Collateral effects of drought and desertification include less irrigated arable land (and increasing dependence on food imports[17]), and a growing flow of the rural population towards urban centres that should host, by 2015, 70% of the region's population. Water scarcity can be regarded from another, more political, angle as well. Various countries, such as Egypt and Syria, largely depend on freshwater resources (rivers) originating from other countries, and are therefore vulnerable to the water policies of their neighbours. Water management frameworks have, however, recently been established in the Maghreb/Mashreq area (e.g. the Nile Basin Initiative and the Turkish-Syrian Joint Water Committee).[18]

A quick review of some basic structural indicators reveals the magnitude of the development challenge with which most MENA countries, including many of the resource-rich countries, will be confronted. That said, failure to address these and other problems lies notably at the political level. From this standpoint, key issues for the future concern the faltering pace of governance reform and democratisation, ethnic and sectarian religious struggles across the Middle East, and the potential for a major systemic crisis in the Mashreq/West Asia.

Three key political challenges

The Mashreq region is in the throes of change as democratisation is widely discussed, distinctive ethnic and sectarian identities are becoming increasingly marked and regimes are toppled (the Baathist regime in Iraq), voted out (Fatah in Palestine) or come under increasing internal and external pressure (Syria, Saudi Arabia, Iran and Lebanon). National power holders perceive these new developments as a source of instability. This is all the more the case, as one has to remember the devastating roles politicised religion and ethnicity have played until now in Afghanistan, Iraq, Turkey and Lebanon. In each case, the results were state breakdown or the loss of state authority over much of its territory. Failed states, lawless territories and rogue regimes throughout the 1990s (especially in Afghanistan and Somalia) were the breeding grounds for extremism and the rear bases for international terrorist networks like Al-Qaida and Mafia-like organised crime networks, mostly drug cartels. But failed states/regions and the harbouring of extremists and criminals are just one element of what has been called the Greater West Asian crisis.[19] The second element is the internationalisation of local conflicts, with the conflict between Israel and Palestine acting as the catalyst and multiplier of regional tensions.

Israel and Palestine

The conflict between Israel and Palestine remains the most important challenge for the international community. Today the question is whether a sovereign Palestinian state will be created on the territory of Mandate Palestine or not. Palestine continues to be the *cause célèbre* of Arab nationalism as well as political Islam and the traditionally anti-Imperialist left. Therefore it will continue to politicise Arabs and Muslims around the world. People in MENA already see all conflicts as local expressions of one and the same conflict: Western hostility towards Islam. Foreign presence in Iraq is largely perceived by people in MENA in the same terms as the Israeli presence in the occupied territories. At the same time, given the geographic centrality of Iraq, a new kind of radical militant underground combining diverse groups from Kashmir to Palestine is developing. It seems likely that, as the chances for a negotiated two states solution look increasingly dim, frustration, embitterment and outright hatred against the West (here understood as Israel, the US and Europe) will intensify, which will in turn aggravate the security situation of EU countries.

Democratisation and islamisation

The intertwined nature of democratisation and Islamisation will pose one of the biggest dilemmas of the next two decades.[20] This apparently odd relationship has deep roots: it is about overcoming both authoritarian power and backwardness. In the twentieth century 'Islamism' emerged as 'political Islam' on the intellectual-political level and generated a more virulent form of 'Islamic fundamentalism' at the societal level.[21] Both were reactions to colonial power-holders and, subsequently, to the misrule of the authoritarian crony-based regimes in MENA[22] and to the dwindling appeal of their nationalist and/or Marxist ideologies. Civil society organisations like the Muslim Brotherhood[23] in Egypt, and later throughout MENA or the Nurcu movement in Turkey, led to Islam becoming a more important factor in society. The aim of these organisations was, and is, to Islamise society from the bottom up. To do so, they resort to a whole variety of social institutions and networks like charities, also engaging in economic activities, and thereby building up so much indirect social pressure that people are constrained to adhere to Islamic norms and mores in public life. This trend is unlikely to abate any time soon: it will remain a cultural as well as a political challenge to the secularised elites of MENA. The degree of confrontation and instability that Islamic activism will cause will depend on the importance that political elites in MENA states attach to secularism.

The approach of political Islam to the issue of democracy has undergone considerable evolution. Initially, promoters of political Islam were very critical of democracy,[24] yet over the 1990s a tendency to support the popular vote emerged. This shift was based on the expectation that, following the successful Islamisation of society, up to 40% of the voters could be mobilised to vote for Islamic parties in many Muslim countries. Hence, the marriage between Islamism and democracy was a marriage of convenience.[25] However, looking ahead to the next decades, democratisation should not be seen as the natural result of this development.[26] Frustration caused by rigged elections[27] or by the international community not accepting the result of elections, risks diluting the attraction of democracy itself. Furthermore, enthusiasm for democracy could also erode as a consequence of political apathy among the masses.

In addition, the relationship between Islamism and democratisation is, and will be in the future, upset by minority extremist movements, which continue to reject democracy as a deviation and a sin against God. These tiny pockets of extremism cannot be included in democratic politics. They have developed a violent approach, for which the term *jihadism*[28] has been coined. It is important to note that a clear inverse correlation is emerging between the maturity of parliamentarism and the appeal of *jihadi* ideologists: the bigger the chance for political Islam to win by free elections, the smaller the spectrum for extremists. Nevertheless, extremists who subscribe to *jihadism* will continue to exist in all Islamic countries and also in Western Europe.[29] By using the tool of terrorism they still have the potential to influence the global debate on how to deal with political Islam. The loosely knit network of global *jihadis* is the main terror threat for the EU, since *jihadis* subscribe to a global agenda and follow totalitarian aims. Their hitherto most famous network is Al-Qaida.[30] On the other hand, Al-Qaida as such is a transient phenomenon, whereas *jihadism* continues to flourish and will possibly convert national, issue-driven Islamic resistance movements into aggressive terrorist organisations.

The fact that this *jihadi/salafi* movement is gathering momentum in most of the MENA countries has three consequences. (1) At the ideological level this means extreme hostility towards and persecution of moderates of any colour, in order to disrupt nascent democratic tendencies, which are perceived as tools of imperialism. (2) At the religious level this has already materialised in aggressive persecution of the Shiites, who are viewed as heretics. But the last (3) consequence has the most disturbing impact not only for the region and the EU, but also for Africa, India and Southeast Asia: trans-border terrorism. As the *jihadi* movement is international – a legacy of Al-Qaida – it can provide rotation of fighters from one conflict to another. Today hundreds of small, independent resistance groups in Iraq provide combat training, while new terrorist tactics originating from Iraq are already in use in Afghanistan and elsewhere. The longer the conflict continues, the more foreign fighters will train and the better and the more sophisticated the post-Al-Qaida networks will be able to operate.

Given the fact that no ideology other than Islamism has serious appeal among the masses in MENA, Islamisation and democratisation will go hand in hand in the region for the foreseeable future. For all their differences, the Turkish AKP, Iraqi UIA, Lebanese Hizbullah and Palestinian Hamas serve as examples. Western support for free and fair elections will be key to promoting democracy in the region. This will include making economic support conditional on good governance, human rights and media freedom – and accepting the results of elections. Needless to say, the democratisation process will be incremental, and is unlikely to result in political regimes resembling Western liberal democratic systems. The establishment of multi-party parliamentary democracy, on the other hand, is an important target *per se*.

For all its shortcomings, this seems the sole viable approach to the de-radicalisation of radical Islamists. On the other hand, this approach will be exposed to considerable challenges. First, new democracies can be badly governed or politically unstable too. Second, *jihadism* will continue to menace both regional regimes and the internal security of the West notwithstanding how democratic, or not, regimes in the region are. Third, radical Islamist forces may decide to use the power that they gained via democratic channels for non-democratic ends. Hence the importance of supporting structural governance reform to sustain the democratic momentum.

Ethnicity and sectarianism

Most borders of MENA states were artificially drawn by the former colonial powers, ignoring ethnic and sectarian realities on the ground. In general, the nationalist elites in MENA countries have always perceived ethnic, tribal and sectarian tensions as national security threats, i.e. threatening national unity and hampering state control over society. MENA states generally did not endorse minority rights but rather tried assimilation, which was temporarily successful during the last century. However, from the late 1980s/1990s onwards, this trend has been reversed and some minorities have openly challenged the state. Currently, minorities are increasingly raising their voices. This is the case, for example, of the Berbers in North Africa,[31] Copts in Egypt or non-Wahhabi Muslims in Saudi Arabia. Two cases however will have a serious impact on regional security: the Kurdish issue and the relationships between Sunnis and Shiites. If these cases are mishandled or tensions exacerbated, the region might live through a phase of instability caused by ethnic and sectarian strife for decades to come.

▶ *The Kurdish issue*

Kurds form sizeable minorities in Turkey, Syria, Iraq and Iran and have often violently fought for their cultural and political rights in each of these countries.[32] The new *de facto* Kurdish state[33] in Iraq serves as a source of inspiration for all Kurds and has the potential to become the nucleus of an irredentist entity. Yet the *de facto* state of 'Kurdistan'[34] will try to resist irredentism as long as it has not gained full independence and will try to find an equilibrium between Turkey, Iran and Syria over the next ten years. Hence, the next decade in this region will be dominated by the 'Kurdish issue' or the 'Kurdish question' i.e. whether *de facto* existing Kurdistan will be recognised as a new state or not. This will increasingly distract diplomatic energy and statecraft from regional countries, the US and the EU and will put the EU's relations with Turkey under stress.

The status of Kurds in other countries, notably Syria, Turkey and Iran[35] will heavily influence the fate of these states since it will transform their ideological foundations. Turkish Kemalism[36] will be hardly recognisable once minority rights for Kurds are fully implemented,[37] and the same can be said for the Baathist ideology of Syria. In either case, the state of minority rights for Kurds can be used as a barometer for the overall situation of democracy.

In the case of Iran, the relative success of the Kurdish nationalist movement(s) is already encouraging other minorities within this country, where only 55% of the population is made up of Persians.[38] Kurds and Azeris have fought for autonomy on several occasions in the twentieth century.[39] Iranian regimes were generally able to keep autonomist tendencies under control by simultaneously suppressing them militarily and proving relaxed on cultural matters, especially regarding the use of minority languages. However, because of widespread frustration with the regime, periphery-centre (i.e. province-Tehran) relations are increasingly regarded from an ethnic perspective.[40] This tendency is exacerbated by outside actors who try to manipulate Iran's multiethnic society in order to undermine the regime. And this in turn seriously affects the policy calculus of Iranian decision-makers, who will be forced to view Iran's multiethnicity from a security-related angle and not from the angle of regional and cultural policy.

▶ *Sunnites and Shiites*
In general, Arab elites to this day perceive Shia Islam as a Persian 'heresy' rather than as genuine Islam. This mindset was, and still is, responsible for the various discriminations which Shias have suffered at the hands of the Sunni elites and population in the Mashreq. With a Shiite-Arab Iraq, it will be less and less possible for Saudi Arabia, Yemen, Bahrain, Qatar, the Emirates and Kuwait to sideline and ignore confessional diversity among its citizens and to deny the existence of Shiite minorities or even majorities.[41] Careful reform is already taking root in Kuwait and Shiites increasingly participate in political life. However, whether the Kuwaiti experience might serve as an example for others depends on the pace of democratisation and whether this would be enough to alleviate sectarian tensions.[42] With some certainty three predictions about the Shias can be made. First, the Sunni elites will no longer be able to ignore the Shias. Second, incremental democratisation might prove a means of easing tensions between different confessional groups. Third, the possibility of Iranian and Iraqi intervention in the Persian Gulf region in support of Shia minorities will remain a source of major concern.

Notes

1. *World Energy Outlook 2005* (International Energy Agency, Paris, 2005), p. 91. It should be noted that some uncertainties exist regarding the validity of these figures, since the increase essentially depends on the Saudis' capacity to increase their own production from 10.5 mb/d in 2005 to 18.2 mb/d in 2030.
2. Regional growth has averaged 5.8% since 2002.
3. During the current oil price boom, only 25% of additional export revenues have been spent (compared to 60% during the 1973 boom). Several countries choose to use additional oil revenues to reduce their external debt. See 'Middle East and North Africa. Economic and Development Prospects 2005: Oil Booms and Revenue Management', Middle East and North African Region, Office of the Chief Economist, The World Bank, 2005. This important report informs some of the considerations developed in this section.
4. Ibid.
5. The private sector represents less than 50% of GDP on average, and is concentrated in a small number of large firms and a large number of micro-enterprises. The public sector provides on average one third of total employment in the region. This ranges from 10% in Morocco to 93% in Kuwait. Ibid.
6. *Global Development Finance: The Development Potential of Surging Capital Flows* (The World Bank, Washington D.C., 2006).
7. Between 2003 and 2004, oil exports accounted for 70 to 80% of the Saudi state revenues. Including gas, energy exports amount to 40% of the GDP of Saudi Arabia and Kuwait, and 40 to 50% of Algeria's. In the case of Iran, oil and gas exports are supposed to represent 40 to 50% of the Iranian government budget. See Energy Information Administration, *Country Analysis Briefs* (http://www.eia.doe.gov/) and the *CIA World Fact Book* (http://www.cia.gov/cia/publications/factbook/geos/ag.html).
8. Some oil-importing countries succeeded in managing their rising energy import bill, with GDP growth rate at 5% in Egypt, 7.2% in Jordan and 4.2% in Tunisia. On the other hand, in 2005, exogenous events reduced growth in Morocco to 1.8% (due to bad climatic conditions for agriculture) and to 1% in Lebanon (due to domestic political tensions).
9. Martin Baldwin-Edwards, *Migration in the Middle East and Mediterranean*, A Regional Study prepared for the Global Commission on International Migration, Mediterranean Migration Observatory, January 2005.
10. In some cases, such as Saudi Arabia, the two trends are connected. On the one hand, Saudi Arabia is planning to reduce the share of its immigrant population to 20% of the current total. On the other, its own population is expected to almost double over the next 20 years, putting enormous pressures on an oil-dependent economy.
11. *Unlocking the Employment Potential in the Middle East and North Africa* (The World Bank, Washington D.C., 2005).
12. Ibid.
13. Since January 2005, with the end of the Multi Fibre Agreement that included quotas for the exports of textile products to developed countries, Tunisia and Morocco have been facing growing competition in the textile sector, a key one for their economies.
14. It is interesting to note, for example, that the non-oil exports of the entire MENA region, are lower than the exports of Finland. See Daniel Muller-Jentsch. 2005, 'Deeper Integration in Trade Services in the Euro-Mediterranean Region: Southern Dimensions of the European Neighborhood Policy', Draft working paper (The World Bank, Washington D.C., 2005).

15. In 2002 Saudi Arabia spent 1.7% of its GDP, roughly US$3.4 billion, in this sector whereas at the same time Egypt invested US$ 750 million on water supply and sanitation services. In the case of Egypt, the water sector – including irrigation – represents 20% of the national state budget. Countries of the Gulf Cooperation Council (GCC) have invested as much as about US$ 4.9 billion in water supply services. See 4th World Water Forum, Middle East and North Africa Regional Document, 16-22 March 2006. Accessible at http://www.worldwaterforum4.org.

16. Demand for water is around 200 km³ per year whereas renewable water resources range around 335 km³/year. To put it bluntly, according to an expert, 'The MENA region's present rate of groundwater withdrawal cannot be considered sustainable over the long run since it far outpaces the rate of replenishment and will relatively rapidly deplete the aquifers'; Mamdouh Nasr, 'Assessing Desertification in the Middle East and North Africa: Policy Implications', in Hans Günter Brauch et al. (eds.), *Security and the Environment in the Mediterranean: Conceptualising Security and Environmental Conflicts* (Springer Verlag, Berlin-Heidelberg, 2003).

17. Food imports already cover more than 50% of the local needs, while 80% of the regional cultivations depend on erratic rainfall.

18. Bilateral or multilateral frameworks for water exploitation are a necessity for some major countries of the region: 97% of Egyptian freshwater resources come from outside and about 70% for Syria.

19. Fred Halliday, *The Middle East in International Relations: Power, Politics and Ideology* (Cambridge University Press, 2005), pp. 154-64.

20. By democratisation we understand the introduction and implementation of parliamentary systems including fair elections and a free press in MENA countries; hence the result of democratisation is parliamentarism, which is not the same as liberal democracy. Regarding 'Islamisation' we suggest, following Günter Seufert, understanding this as all acts and attitudes which aim to endorse the Islamic religion as the very forming power of politics, society and economics. In other words, the term 'Islamisation' might more correctly, albeit rather cumbersomely, be replaced by the coinage 'Islamistisation'. Cf. Günter Seufert, *Politischer Islam in der Türkei. Islamismus als Symbolische Repräsentation einer sich modernisierenden muslimischen Gesellschaft* (Beiruter Texte und Studien 67/Türkische Welten 5, Istanbul/Stuttgart 1997), p. 25, note 1.

21. The standard reference work on future developments in Islamism is: Olivier Roy, *Globalized Islam. The Search for a New Ummah* (Columbia University Press, New York, 2004); this was originally published in French as *L'Islam Mondialisé* (Seuil, Paris, 2002). For an alternative interpretation, see Ahmad S. Mousalli, *Islamic Fundamentalism: Myths and Realities* (Ithaca, Reading, 1998); for a thorough analysis of the problematic use of the terms 'fundamentalism', 'Islamism' etc., see Gilbert Achcar, 'Maxime Rodinson on Islamic "Fundamentalism"', *Middle East Report*, vol. 34, no. 233, Winter 2004, pp. 2-4.

22. Hardly any of the authoritarian rulers in MENA countries were delivering on the economy, since they preferred a state-run system to the free market. The oil-producing countries became what has been described as the 'Rentier State': this term applies to conservative monarchies like Saudi Arabia as well as to revolutionary republics like Iraq under Saddam Hussein. On the use of the term 'rentier state' in the context of regional economics, see Giacomo Luciani, 'Oil and Political Economy in the International Relations of the Middle East,' in: Louise Fawcett (ed.), *International Relations of the Middle East* (Oxford University Press, 2005), pp. 79-104.

23. The standard reference work is still Richard P. Mitchell, *The Society of the Muslim Brotherhood* (Oxford University Press, 1969, republished 1993).

24. On the state of democratisation, see Ghassan Salamé (ed.), *Democracy without Democrats? The Renewal of Politics in the Muslim World* (I.B. Tauris, London-New York, 1994); on the current political debate, see Sigrid Faath (ed.), *Politische und gesellschaftliche Debatten in Nordafrika, Nah- und Mittelost. Inhalte, Träger, Perspektiven* (DOI Mitteilungen 72, Hamburg 2004).

25. On democratisation in the Arab world, see Alan Richards, 'Democracy in the Arab Region: Getting There from Here,' in *Middle East Policy*, vol. XII, no. 2, Summer 2005, pp. 28-36.

26. See for instance the round table discussion 'Democracy: Rising Tide or Mirage?' in *Middle East Policy*, vol. XII, no. 2, Summer 2005, pp. 1-27.

27. See Jillian Schwedler and Laryssa Chomiak, 'And the Winner Is… Authoritarian Elections in the Arab World,' *Middle East Report*, vol. 36, no. 238, Spring 2006, pp. 12-19.

28. Gilles Kepel, *Jihad: Expansion et déclin de l'islamisme* (Gallimard, Paris, 2000); Fawaz A. Gerges, *The Far Enemy. Why Jihad Went Global* (Cambridge University Press, New York, 2005). John L. Esposito, *Unholy Wars: Terror in the Name of Islam* (Oxford University Press, New York 2002).

29. This is the topic explored by Olivier Roy in *Globalized Islam* …, op. cit.; see also the final chapter in Gilles Kepel, *Fitna: Guerre au Cœur de l'Islam* (Gallimard, Paris, 2004), pp. 286-334, which bears the telling title 'La bataille d'Europe'.

30. Among the vast literature on Al-Qaida, the must-reads are : Marc Sageman, *Understanding Terror Networks* (University of Pennsylvania Press, Philadelphia, 2004); Peter Bergen, *Holy War Inc.: Inside the Secret World of Osama bin Laden* (Free Press, New York, 2002); Karen J. Greenberg, *Al Qaeda Now. Understanding Today's Terrorists* (Cambridge University Press, 2005); Gilles Kepel and Jean-Pierre Milleli, *Al-Qaida dans le texte* (Presses Universitaires de France, Paris, 2005).

31. On North Africa, see Paul Silverstein, 'State and Fragmentation in North Africa,' *Middle East Report*, no. 237, Winter 2005, pp. 26-33; on the Berber, see Paul Silverstein and David Crawford, 'Amazigh Activism and the Moroccan State,' in *Middle East Report*, no. 233, Winter 2004, pp. 44-7.

32. Regarding the Kurds, the following authoritative studies are worth consulting: Martin van Bruinessen, *Agha, Sheikh and State. The Social and Political Structures of Kurdistan* (Zed Books, London 1992); Hamit Bozarslan, *La question kurde, États et minorités au Moyen-Orient* (Presses de Sciences Po, Paris, 1997); David McDowall, *A Modern History of the Kurds* (I.B. Tauris, London, 1997)); Martin Strohmeier and Lale Yalçin-Heckmann, *Die Kurden. Geschichte, Politik, Kultur* (C.H.Beck, Munich, 2000); a dated classical work of reference is Gérard Chaliand (ed.), *Les Kurdes et le Kurdistan* (Maspero, Paris, 1978), published in English as *A People without a Country. The Kurds and Kurdistan* (Zed Books, London, 1980, several reprints).

33. On the definition of *de facto* states, see Dov Lynch, *Engaging Eurasia's Separatist States. Unresolved Conflicts and De Facto States* (United States Institute of Peace, Washington D.C., 2004); on Iraqi Kurdistan, see Gareth Stansfield, *Iraqi Kurdistan: Political Development and Emergent Democracy* (Routledge, London, 2003); Brendan O'Leary, John McGarry and Khaled Salih (eds.), *The Future of Kurdistan in Iraq* (University of Pennsylvania Press, Philadelphia, 2005).

34. The geographic definition of Kurdistan is still roughly the same as described in the 16th century history of the Kurds, the *Sharafnâme*; see van Bruinessen, *Agha, Sheikh and State*, op. cit., pp. 11-13; It should be mentioned that there has never been an independent state of Kurdistan with the exception of the short-lived period of the republic of Mahabad in Iran towards the end of World War II.

35. M. Hakan Yavuz and Nihat Ali Özcan, 'The Kurdish Question in Turkey,' in *Middle East Policy*, vol. XIII, no. 1, Spring 2006, pp. 102-119; Henri J. Barkey and Graham Fuller, *Turkey's Kurdish Question* (Rowman and Littlefield, New York, 1998); Robert Olson (ed.), *The Kurdish Nationalist Movement in the 1990s: its Impact on Turkey and the Middle East* (University Press of Kentucky, Lexington, 1996); Robert Olson, *The Kurdish Question and Turkish-Iranian Relations. From World War I to 1998* (Mazda Publishers, Costa Mesa, 1998).
36. On Kemalism, see Hugh Poulton, *Top Hat, Grey Wolf and Crescent: Turkish Nationalism and the Turkish Republic* (New York University Press, New York, 1997), pp. 87-129.
37. On nationalism and language politics in Turkey, see: Hüseyin Sadoglu, *Türkiye'de Ulusçuluk ve Dil Politikalari*, [Nationalism and Language Politics in Turkey] (Bilgi Üniversitesi Yayinlari, Istanbul 2003).
38. See Kaveh Bayat, 'The Ethnic Question in Iran,' in *Middle East Report*, no. 237, Winter 2005, pp. 42-5.
39. On the Azeris, see for instance Jamil Hasanov, *Güney Azärbeyjan: Tehran – Baki – Moskva arasinda, 1939-1945*; [South Azerbaijan between Tehran, Baku and Moscow, 1939-1945] (in Soviet-Azeri), (Baku, 1998); Ali Morâdi Marâghei, *Az zendân-e Rezâ Khân tâ sadr-e ferqe Demokrât-e Azerbayjân (...)*, [From the prison of Reza Khan to chairmanship of the Democrat Party of Azerbaijan] (in Persian), (Nashr-e Awhadi, Tehran 2004).
40. See Bernard Hourcade, 'Iran's internal security challenges,' in Posch, Walter (ed.), *Iranian Challenges*, Chaillot Paper no. 89 (EU Institute for Security Studies, May 2006), pp. 41-59; see also Marie Ladier-Fouladi, 'Population et politique en Iran. De la monarchie à la République islamique', *Les Cahiers de l'INED*, no. 150 (Institut National d'Etudes Démographiques, Paris, 2003).
41. See the Symposium notes 'A Shia Crescent?', in *Middle East Policy*, vol. XII, no. 4, Winter 2004, pp. 1-27; Graham Fuller and Rend Rahim Francke, *The Arab Shi'a. The Forgotten Muslims* (New York, 1999); a general introduction is Heinz Halm, *Die Schia* (Wissenschaftliche Buchgesellschaft, Darmstadt, 1988); Rainer Brunner and Werner Ende (eds.), *The Twelver Shia in Modern Times. Religious Culture and Political History*, (Brill, Leiden, 2001).
42. For a concise analysis of this problem, see Vickie Langohr, 'Experiments in Multi-Ethnic and Multi-Religious Democracy', in *Middle East Report*, no. 237, vol. 35, 4, pp. 4-7.

Sub-Saharan Africa

8

The future of Africa is the subject of two competing narratives: Afro-pessimism versus Afro-optimism. According to the former, Africa will remain a continent of great concern, endemically affected by bad governance, corruption, violence, poverty, droughts or diseases. The optimistic discourse, on the other hand, envisages that new opportunities will emerge, with a new generation of African leaders prepared to confront African challenges and set the stage for the integration of the continent in the globalisation process. Different trends and findings can fuel both narratives. However, stressing the diversity of situations and potential trends within Africa seems more appropriate than imposing a global perception. Above all, the reform of the existing political and economic governance will be a crucial factor in paving the way towards stability and development.

Missed opportunities? Africa's potential and its prospects

Various key statistics highlight the extent to which Africa is marginalised in the globalisation process. It accounts for only 2% of global trade and receives approximately 2% of global Foreign Direct Investments (FDI).[1] Moreover, the poverty rate is already high,[2] while Africa is likely to experience the highest population growth of all continents over the next 20 years (+43 to 48%),[3] even if demographic expansion could be affected by major diseases, such as malaria and HIV.[4] The pace of economic growth will be crucial to alleviate poverty and reach the Millennium Development Goals (MDG). The New Partnership for Africa's Development (NEPAD) vision document outlines that the average growth rate should reach 7% per annum if extreme poverty is to be halved.[5] Achieving this target would require $64 billion of additional fixed investment per annum. The mobilisation of domestic savings, the commitment of foreign donors through Official Development Assistance (ODA),[6] and attracting large flows of FDI, will be necessary to increase investments to such a level. In addition, it is estimated that, depending on the country in question, between 50% and 80% of FDI to Africa is directed to the energy sector.[7] This underlines how limited the diversification of

African economies is, and is likely to remain, in the current governance context.

Lack of investment will lead to major bottlenecks that will significantly undermine growth rates. However, at least three potential sources of African money could provide the bulk of capital for such investment: remittances, repatriation of capital flight and major additional resources from exports of raw materials. First, Africa receives only 5% of the remittances destined to developing countries, amounting to $8 billion, but the real level of remittances, including informal channels of transfer, is believed to be at least 50% higher.[8] Second, as much as corruption,[9] capital flight and limited reinvestments of profits are key issues. An estimated $150 billion has been lost through capital flight and around 40% of African private wealth is located outside the continent.[10]

Third, some African countries will benefit from major growth in their exports of raw materials, as a consequence of global economic growth. Angola's oil production is expected to increase from 930,000 barrels a day in 2003 to 3.28 million in 2020; Nigeria's production could double and reach 4.4 million by 2020.[11] African proven oil reserves, (103 billion barrels) amount to 10% of the world's oil reserves and the size of expected reserves could guarantee long-term growth to the oil-producing country. Aggregated African oil production could theoretically match the foreseen demand even if the absence of intra-continental transportation infrastructure makes this prospect extremely remote. Thus, access to energy will remain one of the main constraints on development. An increase in oil prices is likely to impact negatively on the growth of non-producer countries but the effect on the poorest countries could be less severe than foreseen, since biomass represents the main energy source in countries with a per capita GDP of less than US$300 (in PPP terms).

Beyond oil production, key raw materials such as iron, copper and precious metals, are also very likely to lead to an increase of FDI in specific countries.[12] On the other hand, major disparities among African countries could stem from international investment and cash targeting only those rich in raw materials. The rising price of oil and other raw materials is indeed detrimental to non resource-rich countries. It has induced major current account vulnerabilities in other countries, whose energy import bill has significantly increased. In 2005, oil-exporting countries experienced a rising trade surplus (19.8% of GDP) whereas non-oil exporting countries experienced a deepening of their deficit (6.6% of GDP).[13] Growth is also unevenly distributed in the region: oil-exporting countries posted a strong 6.4% growth in 2005 while it was 4.3% in small oil-importing countries.[14]

Turning to other key sectors of the economy, agriculture and manufacturing will face constant major challenges. Agriculture in Africa is currently under-exploited, in spite of its significant potential. In most of Africa, agricultural productivity remains low and has not kept pace with population increase. Even if droughts are a recurrent issue in the Sahelian Belt and the Horn of Africa, arable land is largely available, as Africa has the lowest density of population in the world.[15] However, only 7% of arable land is currently irrigated. The trend is negative, as the irrigation rate has fallen since the 1980s. Hence, imports of food – cereals, rice, as well as meat,– are expected to significantly increase.[16] Concerning outward-looking agriculture, major hindrances are frequent: access to foreign markets, the frequent lack of priority attached to the agriculture sector, the vulnerability of over-specialised agriculture, etc. Even if it were possible to achieve major improvements in the agriculture sector, intensified agricultural productivity would result in a pool of labour being made available. If alternative job opportunities were not created, additional unemployment and social tensions could ensue.

The diversification of the African economy remains a crucial issue, in particular the development of the manufacturing sector. If unaddressed, various economic bottlenecks will continue to hamper its development: lack of investment, poor infrastructures, limited internal markets, etc. However, textile manufactures have provided a significant example of successful labour-intensive production in numerous African countries.[17] Nevertheless, a sustainable success remains unlikely. Economies of scale and narrowing trade preferences favour producers of other continents, in particular Asia, regarding labour-intensive sectors. In addition to global competition, political and economic governance in Africa could remain a major constraint to the development of African manufacturing.[18]

Political and economic governance: the crucial factor

In the absence of major fixed investments and reforms of the economic environment, diversification of the African economies is unlikely in the next twenty years. Hence, African economies will remain dependent upon exports of raw materials, as in past decades, and growth will be constrained by the size of reserves and international prices. These various challenges and opportunities will be highly affected by the political and economic governance of African rulers. If governance is not adapted, best-case scenarios are unlikely for 2025. Governance in Africa is often characterised by two intertwined elements: limited – or even virtually non-existent – political checks and balances and a rent economy that often precludes the diversification of economic activities.

Most African economic governance not only hinders economic diversification, but also provides a breeding ground for violence. Most African countries rely on a rent economy with a key role reserved for the state apparatus.[19] Typically, rulers do not rely on taxes but on the international prices of raw materials. Moreover, the legacy of authoritarian regimes will continue to affect the development of the private sector. In the vast majority of authoritarian regimes during the Cold War, an autonomous private sector was often perceived as a political threat. Two kinds of entrepreneurs have hence dominated the African economic landscape: foreigners and cronies of the regimes. Foreigners, including both nationals of former colonial powers and others (such as Lebanese, Indians or Greeks), have had a key economic role. To a lesser extent than cronies of regimes, they have often benefited from limited competition, even if they were occasionally expelled.[20] As a consequence, African companies are often poorly equipped to compete in a globalised world. However, significant exceptions stress a potential dynamism that could contribute to economic growth. These include South African telecommunications companies and African traders connecting their markets to Dubai or Hong-Kong.[21]

From a political standpoint, even if elections have become commonplace in numerous African countries since the end of the Cold War, this has not led to significant check and balance systems. Independence of the judiciary remains the exception rather than the rule;[22] corruption is a major obstacle to rule of law and strongly affects economic environments. Presidencies and governments most often overshadow parliaments. Moreover, both rulers and opposition often share a common 'winner-takes-all' approach to elections. Changeover of political power between political parties remains a sensitive process that often leads to instability. Maintaining the freedom of the press is often a daily struggle, even if various dynamic news groups exist in Southern and Eastern Africa, as well as in Senegal. Social movements have been weakened in the wake of the early 1990s repressions and economic crises,[23] often paving the way to more violent conflicts, and even armed rebellions.[24] A lack of reforms and progress on these issues means that numerous conflicts in Africa are likely to continue. Moreover, they may acquire an increasingly regional dimension, as more and more conflicts spill over borders. In the foreseeable future, improvements can only be incremental and governance inadequacies are likely to remain a major source of concern and instability.

African rulers and foreign powers

Changes in relations between African and foreign rulers will be a key factor. African rulers have often proven very effective in adapting to foreign demands and discourses while preserving existing governance systems.[25] They have hence become highly skilled at exploiting competition among foreign players and in neutralising conditionalities that do not fit in with their own goals. The renewed interest in African raw materials could reinforce such a situation. This coincides with the economic exapnsion of emerging economic powers, such as China, India or Brazil. Chinese and Indian FDI in Africa will continue to significantly increase, as in the past five years. For emerging economies, securing access to natural resources is a key issue to sustain high growth rates. Africa will continue to provide key opportunities for such policies. Competition with western firms is often less intense, reserves of natural resources are very large and, in the case of the Chinese regime, it can use its political support for 'rogue states'[26] as an additional leverage.

This trend could provide additional resources to maintain non-democratic regimes and develop their security forces.[27] In addition to exports of raw materials, ODA has often been siphoned off locally in order to fund an authoritarian state apparatus[28] and support clientelist practices. By usually focusing on African rulers, foreign players have often become disconnected from domestic constituencies that support reforms. Enlarging support to efficient African NGOs or competitive entrepreneurs will remain a major challenge to obtain sustainable changes in governance. Nevertheless, both the use of ODA and increasing revenues from raw materials exports underline how crucial the role of African rulers will be to improve stability and reduce poverty.

The global terror issue could provide an additional umbrella to obtain resources from foreign powers to reinforce security or military capabilities.[29] Along with oil, the fight against terrorism has led to a partial revival of US interest in the African continent. However, various failed military interventions are likely to hamper the direct deployment of troops. Similarly, former European colonial powers will continue to downsize their direct military presence. A policy of intervention by proxy – in particular via the United Nations and the African Union – is thus likely to be maintained or even developed. However, new alliances are also likely to appear in the wake of African policies implemented by new global players such as China, India and Brazil. In spite of their weaknesses, the Sub-Saharan states represent 44 seats in multilateral arenas: a 'south-south' axis could gradually appear, in which African states could provide a 'voting bloc' to emerging powers.

Other drivers of conflict

A lack of change in governance is likely to fuel conflicts. Although the number of conflicts and the level of violence are not higher than in other continents, the image of Africa is associated with endemic violence.[30] Moreover, rent economies associated with nepotism and corruption often provide a breeding ground for conflicts. Economic and social grievances are exacerbated by the absence of economic redistribution. The latter is often interpreted in communitarian terms and fuels tensions between communities. This adds to the ongoing problem of xenophobia on the African continent which is very likely to continue to prevail over the next twenty years. Xenophobia, either against foreign nationals or among communities within a country, is a widespread phenomenon in Africa.[31] Fuelling local tensions is part of many African rulers' strategies to preserve their power.[32] As conflicts in one country often spill over to neighbouring countries, such situations are likely to pose a major threat to stabilisation processes.

Three centres of instability are likely to remain or even to reach a renewed intensity of violence in West Africa, the Great Lakes region and Central Africa. Although apparent improvements have occurred in Sierra-Leone and Liberia, Ivory Coast and Guinea are only beginning to enter into conflict dynamics. A deteriorating situation, especially in Guinea, in the coming years is likely to have repercussions for Sierra-Leone and Liberia. Similarly, in the Great Lakes region (Uganda, Rwanda, Burundi, DR-Congo), conflicts are rotating from one country to another rather than being solved. An emerging 'triangle' of conflicts (Darfur-Chad-The Central African Republic) constitutes an additional major source of concern. Flows of weapons and widespread violence are very likely to continue to prevail, even if intemittent improvements may be observed. Moreover, the unity of Sudan will be at stake in 2011, as a referendum in South Sudan will decide whether this part of Sudan will secede or not.

In these three main areas of instability, the conflicts mainly stem from brutal political and economic governance, as described above. These conflicts also draw attention to a key aspect of African conflicts: the stronger the state, the more extensive the violence.[33]

Southern Africa could remain the most stable and promising part of Africa. However, internal developments in Zimbabwe could have destabilising regional implications. The serious economic and political deterioration that has taken place in this country is unlikely to be redressed by major improvements in the coming years. Flows of migrants from Zimbabwe to neighbouring countries are likely to continue and internal violence could resume, as happened in the 1980s. In West Africa, Nigeria is the object of increasing attention regarding its internal stability. However, in the past this country has shown little tolerance for secessionist movements, which at times were brutally repressed. As with any conflict or potential conflict in Africa, improving governance must remain at the core of any approach to containing violence.

Notes

1. i.e. $18 billion in 2005. *Global Development Finance: The Development Potential of Surging Capital Flows* (The World Bank, Washington D.C., 2006).
2. In 2005, the percentage of African population living with less than $1 per day increased to a historical high of 46%. This statistic can be highly questioned but provides an element of comparison with other continents.
3. According to World Bank or United Nations forecasts, it is estimated to grow from 731.5 million (11.4% of the global population) today to more than 1 billion in 2025 (12.6% of the global population).
4. Malaria is currently the most lethal disease in Africa. Cases of malaria could increase or decrease over the next 20 years, depending on the efficiency of medicine and public health policies. HIV infects 9% of the total African population; this represents 70% of global HIV infection. However, epidemiologists tend to consider that the epidemic has reached its peak in Africa, while its spread has just started in Asia – in particular in India and China.
5. In 2005, the average growth rate in Africa was 5.8%. However, this growth mainly stems from a renewed interest in African raw materials.
6. Currently, ODA to Africa amounts to $26 billion, including debt relief (worth $5 billion for Nigeria only) and emergency and reconstruction grants.
7. World Investment Report (United Nations Conference on Trade And Development [UNCTAD], 2004); *Economic Development in Africa, Rethinking the Role of Foreign Direct Investment* (UNCTAD, 2005).
8. *Global Economic Prospects 2006: Economic Implications of Remittances and Migration* (The World Bank, 2006). It should be noted that the figures reported by the World Bank include workers' remittances, compensation of employees and migrants' transfers. These figures, both for Africa and for the world, are double the figures given in the estimates of the *World Economic and Social Survey: International Migration* (United Nations, 2004). See also Cerstin Sander and Samuel Munzele Maimbo, 'Migrant Labor Remittances in Africa: Reducing Obstacles to Developmental Contributions', *Africa Region Working Paper Series* no. 64 (The World Bank, Washington D.C., 2003). According to the authors, if unrecorded flows are added, remittances to Sub-Saharan Africa could be worth $10 billion.
9. Which is not significantly higher in Africa, compared to numerous Asian and Latin American countries. See for instance *Global Corruption Report 2006*, Transparency International, 2006. Available online at: http://www.transparency.org/publications/gcr.
10. The Africa in 2020 Workshop, US National Intelligence Council, 16 March 2004.
11. African oil producers already represent 28% of Chinese imports and 15% of American imports. The latter could increase to 25% by 2015.
12. Such as Gabon or Liberia for iron, Zambia or DR-Congo for copper.
13. *African Economic Outlook 2005/2006* (OECD, Paris, 2006).
14. *Global Development Finance ...*, op. cit.
15. 44 inhabitants per square kilometre in 2004.
16. Imports of cereals are expected to triple 25 years from now.
17. Lesotho, Madagascar, Kenya, Namibia, Nigeria, etc.
18. As an example, the 'Look East policy' of Mugabe in Zimbabwe favours Chinese imports of commodities and is already leading to a collapse in Zimbabwean manufacturing.
19. Within the state apparatus, security forces and intelligence services guarantee the con-

trol of the rent economy to the benefit of the rulers. This is the case of Angola, Sudan, Congo, Chad, Equatorial Guinea, and Nigeria with oil; Guinea, DR-Congo and Gabon with minerals; Ivory Coast and Rwanda with agricultural exports, etc.

20. As happened in Uganda with Indians under Idi Amin Dada, Belgians under Mobutu in Zaire or, more recently, white Zimbabweans.
21. In particular for textile commodities. Such dynamism can also occur with entrepreneurs in war-torn countries and reinvestments of profits, as in Eastern DR-Congo.
22. South Africa is a notable exception. Zimbabwe's independent judiciary was eventually subjugated by the Presidency.
23. South Africa is a major exception with the COSATU trade unions, although they remain linked to the ANC, the hegemonically ruling party.
24. As illustrated by the evolution of Northern Sudan, Ethiopia, Ivory Coast, Democratic Republic of Congo, the oil-producing region of Nigeria, etc.
25. This is an age-old problem. At the beginning of colonisation, colonial troops were often used as a proxy in African power struggles; during the Cold War, alliances and shifts of alliance with the Western world and the USSR provided resources to establish authoritarian regimes in most African countries. In the late 1980s, economic reforms and privatisations were often turned into an instrument for local power struggles (as in Ivory Coast for the cocoa sector), as well as reforms of governance since the 1990s.
26. Such as Sudan or Zimbabwe.
27. Which is already occurring in various places including DR-Congo, Chad and Sudan.
28. Cf. inter alia James Ferguson, *The Anti-Politics Machine, 'development', depoliticization and bureaucratic power in Lesotho* (University of Minneapolis Press, Minneapolis, 1994) and Peter Uvin, *Aiding Violence: The Development Entreprise in Rwanda* (Kumarian Press, West Hartford, 1998).
29. Money laundering, in particular, might become an increasingly problematic issue in Africa.
30. If related to the same demographic scale, the number of casualties in Ivory Coast is lower than in Southern Thailand. Ituri, the most violent area in DR-Congo, has a population similar to that of Bosnia and conflicts have been of similar duration so far. However, the number of casualties is three to four times lower.
31. As illustrated by Zimbabwe and policies targeting white nationals. Similar processes have become more and more obvious against white French nationals in West Africa, or against Chinese nationals where their presence has recently increased. Angola is a major exception in avoiding racial tensions. Between African nationals in provinces, there is frequently a divide opposing 'natives' of a province and 'non-natives'.
32. As in DR-Congo (Katanga, Kivu), Sudan (Darfur, Eastern and Southern Sudan), Ivory Coast (Western part of the country), Zimbabwe and its 'racial divide', etc.
33. As illustrated by the genocide in Rwanda in 1994, which could not have been carried out without highly efficient administrative structures.

The United States

The United States will remain the world's leading superpower for the foreseeable future, especially when its status is measured by the classical indicators of economic and military power. However, to what extent the US will succeed in preserving the hegemonic position it has enjoyed since the end of the Cold War is an open question. Exceptionally for the developed world, the population of the US will increase by over 17%, reaching 364 million by 2030, and the major contribution to this increase will be from Hispanic immigrants. The US economy will continue to grow at respectable rates (on average 3% annually) thanks to America's leading role in technological innovation, and research and development. US expenditure on R&D amounts to roughly one third of the world total[1] and, with 37 of the world's top 50 universities located in America[2] and an unparalleled inflow of scientists from all over the world[3], the US looks set to maintain its competitive edge in the knowledge-based economy in the coming years.

Between 1997 and 2004, the US was the world's economic engine, accounting for 46.6% of the increase of the global aggregate demand.[4] However, with its unprecedented current account deficit nearing $800 billion, or 6.4% of GDP, and a budget deficit that went from a surplus of 1.3% of GDP in 2000 to a deficit of 4.1% in 2005, the US economy may be subjected to periodic fluctuations with negative global implications. There is no doubt that the stability of US finances is now dangerously dependent on external sources: the US must import about $1 trillion of foreign capital per year to finance its deficit and its own FDI. Solutions to this state of affairs will likely include a reduction of the US budget deficit, the realignment of exchange rates and the expansion of domestic demand in other major economies.[5]

Culture, society and politics

The traditional image of American society conveys the notions of openness, small government, multiculturalism and a secular state. Some of these traits have been debated in recent years – for example, America's constitutional secularism. Some have even questioned whether America has not evolved towards becoming a western theocracy.[6] However, despite these controversies, opinion polls confirm that America remains attached to its traditional values, including state secularism.[7] On the other hand, it is increasingly evident that some traditional tenets of 'Americanism' are being challenged and are likely to change over the next 20 years. Four aspects of American culture and society deserve particular attention in this context. These are: the impact of Hispanic immigration; inequality; the growing role of the state; and the place of religion in US politics.

A more Hispanic America

America is a country of immigrants and multiculturalism is its defining feature. In the past America successfully absorbed huge waves of migrants, primarily from Europe, with no serious challenge to its political system and its core values. However, the ever-growing wave of Hispanic migrants, especially from Mexico, is beginning to change the country's social make-up, gradually transforming it into a bilingual society, putting strains on the welfare system and changing attitudes towards migrants. The current wave of Hispanic migration is different from those that occurred in the past for various reasons. Most important among these are its sheer scale and the illegal nature of the immigration, its dimensions of regional concentration and proximity to the migrants' country of origin, historical presence and ambivalent assimilation.

In 2004 the Hispanic population represented between 13-14% of the entire population. This figure is projected to grow to 20% by 2030 and to 25% by 2050. There is a clear and growing dominance of Mexicans within the migration community. In 2004 Mexicans constituted almost one-third (31%) of the total foreign-born population in the US, a figure which has more than doubled since 1990. A vast majority of Mexicans enter the country illegally, around 400-500,000 annually. The total number of illegal migrants in the US was estimated to stand at 10.3 million in 2004.[8]

The sheer number of Hispanic immigrants, although spectacular, is not in itself unprecedented. America experienced comparable waves of migration, especially from Germany and Ireland, in the late nineteenth and early twentieth century. However, it is the first time in its history that over half of those entering the US speak a single language which is not English. Moreover, English language acquisition amongst the Hispanic and especially Mexican, immigrants is slow and reveals ambivalent attitudes.[9] Mexicans of all generations also lag behind other immigrants in terms of their economic status and education. Almost 70% of the first generation do not have a high school diploma and most of them live below the poverty line. This situation improves from generation to generation but much more slowly than is the case for other ethnic groups.

The size of the Mexican community is certainly one of the major factors holding back its integration, but there are also other issues at work, not least, the regional concentration of the migrants and their proximity to Mexico. The vast majority of Mexicans settle in the South West along the border with Mexico — in California, New Mexico, Arizona and Texas — where they remain within immediate reach of their country of origin. Proximity to families and friends on the other side of the border makes it easier to opt out of the integration process. Another factor militating against integration is emerging Mexican nationalism, based on the claim that the US Southern border areas used to belong to Mexico.

Samuel Huntington has argued that, despite its multiethnic and multicultural make-up, America's values are essentially Anglo-Protestant at their core.[10] Whether this claim is accurate or not, there is little doubt that the continuing influx of the Hispanic population will erode the prominence of Anglo-Protestant culture and bring considerable changes to American society. Perhaps most importantly, the US will effectively, if not officially, become a bilingual society. From a political standpoint, Hispanic migrants show a marked propensity to demand more robust welfare provisions and a stronger role of the state. Indeed, for very different reasons, demands for a more ubiquitous state may also come from the non-immigrant sections of the population.

The role of the state

In the traditional American model of state-society relations, the state is meant to act as a *'night watchman'*. It should make sure that people can enjoy their rights, pursue self-enrichment and that they don't kill each other, but the state should not overstep its role and interfere in people's lives. Today, Americans are still suspicious of the state and more critical of its ability to improve their lives than is the case with Europeans.[11] In other words, the night watchman paradigm still holds true in America.

However, to perform its role as the protector of citizens in the age of terrorism, the 'watchman' has become ever more intrusive. The events of 9/11 had an overwhelming impact on the mindset of US citizens leading to an incremental evolution in their perception of the role of the state. The security services were revamped and a powerful Department of Homeland Security was created. The government has regularly introduced new emergency measures that at best caused inconvenience and at worst circumvented individual freedoms. The President has admitted authorising eavesdropping on personal telephone conversations on a massive scale. All of these sweeping changes came with the nation's approval, suggesting that Americans are increasingly prepared to accept intrusion into their lives in order to enhance their security. This could mean the beginning of a change in the night watchman paradigm. An important question for the future is, therefore, whether the Americans' acceptance of the more intrusive state will last beyond the immediate impact of 9/11.

A call for a stronger state is also coming about in the context of immigration reform. The key issue with tackling illegal migration is the state's inability to trace its over 10 million unauthorised immigrants or even prevent them from taking up employment. In order to strengthen the state's ability to enforce the rule of law, the President and both Houses of the Congress have debated immigration reforms that increase the number of border guards, introduce the system of biometric work permits and provide measures for the policing of illegal employment.[12] None of these proposals would be particularly new and dramatic for Europe, but for America they represent a significant shift in the perception of the state as the central actor and as a solution to the problem.

Should America experience another terrorist attack on its territory, there is no doubt that the nation's permission for an ever more intrusive role of the state will grow further. This does not necessarily mean that America will move away from its core values (especially relating to economic freedoms) but some individual rights and civil liberties are likely to be subject to more restrictive interpretation whilst the role of the government will grow.

Inequality

The US will become not only a more diverse society, due to growing migratory flows, but also, if current trends continue, a more unequal one. The US exhibits a high level of income inequality for a developed country. The gap between rich and poor began to widen at the beginning of the 1980s, with the difference between bottom- and middle-income sections of the population growing. Since 2000, the median worker has experienced only a slight 1% real wage rise (as against 6% in the 1995-2000 period). In contrast, the share of aggregate income going to the richest has boomed in the 1980-2004 period: it has doubled for the top 1% share (from 8% to 16%), tripled for the top 0.1% (from 2% to 7%), and quadrupled for the top 0.01% (from 0.65% to 2.87%).[13]

Inequalities are even more marked concerning wealth distribution. One third belongs to the richest 1%, one third to the next 9% and the remaining third to the other 90%.[14] At the same time, however, traditional American values associated with economic freedom and individual entrepreneurship endure. Americans largely believe that success is within reach for the individual and continue to regard positively the opportunities offered by a highly flexible and dynamic economic model.

Religion and politics

American society is deeply religious, indeed in many ways pious. 94% of Americans believe in God and a clear majority is of the view that their country is not sufficiently religious. One in three Americans believes the Bible is the actual word of God and that it should be taken literally. A majority believes that life on earth was created or evolved under the guidance of a Supreme Being. When compared with the rest of the world, some features of American religiosity place them close to the Muslim world, Africa or Latin America but sets them apart from the majority of European nations.[15]

Although the US is constitutionally a secular republic, religion is an important factor in American politics. It has been argued that the Christian conservatives have been on the offensive in recent years, with their influence having a tangible impact on politics and undermining the constitutional separation of the state and religion. President Bush, himself a newborn Christian, has made no secret that his agenda is often driven by his faith. Many members of the Congress are devout Christians as are some members of the Supreme Court. Personal beliefs and attitudes towards abortion were a major factor during the recent debates regarding appointments to the Supreme Court. In fact, the current conservative majority on the Court may open the way for the overruling of the *Roe v. Wade* 1973 decision that legalised abortion.

However, despite these developments, it is premature to conclude that the religiously-inspired agenda will progress further and will deepen its influence on US politics. Although religion will undoubtedly retain a high influence on the social beliefs of Americans, the nation has not changed its views on the desirability of separating religion from day-to-day politics. For example, a plurality of Americans is of the view that organised religion has too much influence on the country and almost three quarters oppose the clergy's involvement in political campaigning in churches. Americans also continue to have a pragmatic approach to religious issues, as evidenced by the fact that most of them support stem cell research and a woman's right to abortion.[16]

As regards foreign policy, there is some evidence that the nation's religiosity has an impact on Americans' views on Israel. Over a third of Americans believe that Israel fulfils a Biblical prophecy and three times more Americans sympathise with the Israelis rather than the Palestinians.[17] It has been also evident for some time that the Christian conservative groups have become the staunchest supporters of Israel and especially of the Likud party. However, beyond this point the evidence of the impact of religion on foreign policy is negative or inconclusive. A vast majority of Americans regard their religion as a private matter and express no interest in imposing their views on others.

There is no sign that America's religiosity will abate in the future. But there is also no clear indication that the impact of spiritual beliefs on public life will grow. In fact, opinion polls suggest the opposite. The impact of religious beliefs on views having a bearing on foreign policy will remain marginal.

Foreign policy challenges

9/11 and the War in Iraq

The terrorist attack on the 11 September 2001 and America's subsequent reaction to it, especially its invasion of Iraq, undoubtedly represents one of the major critical junctures in US foreign policy. 9/11 came at a time when America was enjoying a sense of growing international confidence. The foreign policy of Bill Clinton was marked by continuity with priorities that often originated from the Cold War era (e.g., the Balkans, stabilising Russia, enlarging NATO) and a co-operative style of leadership (engaging China, negotiating with North Korea, investing in the Middle East peace process). The arrival of George W. Bush coincided with the rise of a neo-conservative agenda that advocated a more unilateral, more confident and more militaristic approach. Basically, the neo-conservatives argued that America had the means to change the world and that this is precisely what it should be doing.[18] Yet, at the beginning of Bush's presidency it was by no means clear or even likely that the neo-conservative agenda would be adopted by the administration and the American people seemed to be in no rush to see their troops deployed in nation-building and democracy-promotion around the world.

However, the events of 9/11 changed all that. The threat to America came from sources that arguably could not be contained by traditional diplomacy and appeared to be linked to regional instability, religious fanaticism, authoritarian politics and failed statehood. As a result, the neo-conservative argument of 'going out there and changing the world before it can threaten you' prevailed. Consequently, the administration adopted the doctrine of pre-emption, which has been supported by two-thirds of Americans,[19] and has been applied to the invasion of Iraq.

Over three years after the intervention in Iraq, the failure of this policy, and with it that of the philosophy that underpinned the war, is apparent. The official purpose of the war was to enhance America's security and to build a democratic, stable and pro-western Iraq. In many respects the opposite was achieved: instability and insurgency are now raging in Iraq. 2,500 Americans have already died in the conflict and by June 2006 over $320 billion of taxpayers' money had been spent on the war. In the meantime the US lost the few friends it had in the region and its image and international respect has plummeted to an all-time low.[20] The war is also becoming increasingly unpopular at home, with a majority of Americans being of the view that it was a mistake, for which they blame President Bush – as reflected in the fall in his popularity ratings from 85% in the aftermath of 9/11 to 35% in June 2006.[21]

The failure in Iraq has signalled the end of the 'neoconservative moment', yet the question remains: what will it be replaced by? This depends both on the persistence of current threats (terrorism, proliferation, failing states, etc.) and the nature of the new ones. It also depends on the resonance in the US of the 'regime-change' agenda, with a focus on the nature of the political system of other countries as a key factor in deciding on interventions therein. Yet another important consideration resides in the increasing awareness of the limitations of the US's resources and the waning attraction of America's soft power.

The waning attraction of America's soft power

Nowhere was the impact of the war in Iraq as evident as in the altered global perceptions of America. Between the events of 9/11 and 2005, America lost friends literally everywhere – with favourable perceptions of the country declining by half and more in France, Germany, Indonesia, Morocco and Turkey. Even in the traditionally pro-US Britain, Netherlands and Poland the positive image of the US suffered considerably. Moreover, America is not only disliked but also feared. For example, in 2006 a larger proportion of Western Europeans considered the US to be a greater threat to world peace than North Korea and Iran.[22]

Whilst there is nothing new about anti-Americanism, the post-Iraq phenomenon appears to be more enduring and more consequential in terms of policy than was the case historically. In the past, there were predominantly two sources of anti-Americanism: the first was based on the resentment of America's overweening cultural influence and the second was fuelled by a negative view of American foreign policy during the Cold War. However, in the past none of these negative views affected the perception of the US as the crucial element of international stability. Nor did they affect attitudes to the American people, who remained largely liked and respected around the globe.[23] In addition, for many foreigners the US remained the model they wanted to emulate and America was seen as 'the land of opportunity'.

The post-Iraq rise of anti-US sentiment is both deeper and broader. In the past, anti-Americanism was chiefly apparent among the older sections of populations. The impact of the war in Iraq has been such that that it is the young generations that appear to have become most virulently anti-American. For example, in 2005 the younger generations in France, Spain, Germany and the Netherlands held the strongest anti-US sentiments of any age group in these countries. Anti-US sentiment doubled among the youth of traditionally pro-American Britain and Poland. Significantly, with the exception of India, America is also no longer seen as the land of opportunity with Australia, Canada, Great Britain and Germany all being more popular destinations for global youth.[24]

From the US point of view it is of course worrying that the generation of the world's future leaders is at best unimpressed by the US and at worst actively dislikes it. The other important implication of this phenomenon is a resentment of unipolarity and the popular desire for a more balanced world. For example, the aftermath of the war in Iraq saw a surge in support for a stronger and more independent foreign policy role of the EU, which, again, was especially apparent among the younger generations. China, the most likely future competitor of the US, began to be perceived more favourably around the world. In fact, in most European nations, even in the UK, China has had a better image than the US.[25]

The boosting of the State Department's Public Diplomacy divisions and the underlining of the normative aspects of US foreign policy, e.g. the prominence of promoting democracy in the 2006 Security Strategy, suggest that the administration is aware that it has a big image problem on its hands and that it is investing in addressing it. However, none of these changes indicate that the administration is prepared to fundamentally alter its policies; rather, it seems intent on carrying on as before while at the same time investing in appearing more agreeable and attractive. This is unlikely to change any time soon and the reason for this is not just the policies of the current administration but the views of the American public, which often remain at variance with mainstream views around the globe. On all the major global governance issues, such as the Kyoto protocol, pre-emption, the role of the UN and international aid and welfare, Americans are at odds with global public opinion. As long as these differences remain unbridgeable the decline in America's soft power is likely to continue.

Dealing with rising China

There is no doubt that the evolution of US-China relations will be of major importance for the future of the global order. No other power challenges America's global prominence to the extent that China is likely to do. The rise of China has already altered the balance of power in East Asia and America's prominence in the area. China's growing economic presence in Africa, the Middle East and Latin America has provided an alternative to the US's (and the EU's) influence there and has weakened its leverage *vis-à-vis* the regimes with which it has strained relations, such as Iran, Sudan and Venezuela.[26]

However, although the rise of China preoccupies American diplomats and the Pentagon planners, it would be premature to assume that this relationship is bound to grow more acrimonious. Sino-American relations are just too complex and too multi-layered to judge by exclusively focusing on regional security. On the positive side, there is America's role in bringing China back into the global system and close societal links between the two countries. For example, a considerable proportion of China's young and upcoming elite has been educated at American universities. On the other hand, the two states are locked in conflict over Taiwan and they increasingly compete for influence in the wider arena of East Asia, creating some nascent elements of balance-of-power politics there.

The future of the relationship will be first and foremost defined by the evolution of China itself and its foreign policy posture. The top issues will concern the evolution of China's political system and its behaviour in the global and regional context. But much will also depend on the US. Although 'China hawks' are very vocal in the US, they are not prominent in the current administration.[27] In the event of them gaining the upper hand and real influence, a cornered China may respond by building regional alliances and openly working towards excluding the US from Asia. This would produce a counter-reaction from the US, possibly leading to a regional Cold War.

However, as long as China is evolving without threatening the US it is unlikely that American public opinion would rally behind a policy of confrontation. There are several reasons for this: the two countries are in fact very closely interlinked economically – America buys huge amounts of China's exports whilst China keeps financing America's budget deficit by buying its bonds and currency. Although not always easy, on the whole China has not been uncooperative in the international context. It has rarely explicitly objected to the US position on the UN Security Council, its role in dealing with North Korea is seen as constructive, and it has avoided confrontation with the US outside its own regional context.

The US will undoubtedly strive to retain its influence in East Asia. It will work closely with Japan and it will remain committed to the defence of Taiwan (though the extent of this commitment will depend on Taiwan itself). However, as long as China evolves in a way that is perceived by its neighbours as unthreatening, it is inevitable that the US's position in the region will be challenged and perhaps progressively marginalised over time.

Energy security

On a structural and strategic level, US foreign and security policy will be directed to ensuring the steady flow of energy supplies. The US will be the primary consumer of energy in the world for the next twenty years and beyond. US energy demand is expected to grow faster than in any other OECD country, by about 36% between 2002 and 2025. Technological and policy options have been outlined to enhance energy efficiency in production and consumption, and to invest in renewables, but these are unlikely to match growing energy demand. By 2030, it is expected that America will need to import 66% of its oil (as opposed to 47% today) and around 20% of its gas (4% today) from abroad. In addition, given decreasing gas production in Mexico and Canada, US demand for LNG might rise steeply, with a strong impact on an already strained market.

Unlike Europe, the US is not relying on any specific global region for its imports,[28] but some regions will undoubtedly count more than others in the years to come. As stated in President Bush's 2006 State of the Union address, imports from the Middle East should decrease by 75% in the next twenty years.[29] On the other hand, the strategic importance of Africa and Latin America for US supplies will only grow. Africa is expected to account for 25% of US oil imports in 2025. Nigeria, already providing 10% of US oil supplies and expected to become a large source of LNG, will continue to play a pivotal role.[30] Consistently, Nigeria and other oil-rich African countries like Angola, are receiving the bulk of US investment, and civil and military aid in the region.[31] As far as South America is concerned, the US is increasingly worried about the recent trend of re-nationalisation of energy resources, and a debate has been launched on how to react to this.

Approaches and behaviour

Depending on how the factors outlined above will pan out, US foreign policy is likely to be guided by one of the two following ideological options, or a mix of both:

Realistic Wilsonianism – This approach represents a fusion of liberal internationalism and realism. It is based on the assumption that international institutions are indispensable in the age of globalisation and that the US should promote them. Although sceptical about the value of the UN framework, this approach stresses the role of functional organisations such as the WTO as well as disarmament or environmental regimes. The liberal assumption that internal arrangements in other states are important for their external behaviour will remain one of the guiding principles of American foreign policy and the US will differentiate between democracies and authoritarian states. The US will promote democracy and human rights but it will not pursue 'regime change' by military or other coercive means. The underlining principle of this approach is that a genuine change could be aided from the outside but ultimately it can only originate from within the states and societies themselves. Military interventions would be restricted to self-defence and preventing humanitarian disasters (limited to the prevention of genocide).[32]

Neo-Jacksonianism – This trend would entail a return to a less active foreign policy. This ideology could gain prominence following America's success with protecting the homeland coupled with the growing resentment of its engagement in Iraq, Afghanistan and elsewhere. The Neo-Jacksonian security policy would concentrate on domestic defence – homeland protection and developing a viable missile defence system. US forces would be withdrawn from around the world with the exception of permanent, but downsized, bases in Europe, the Asia Pacific region and probably the Middle East. The US would limit its engagement in multilateral organisations, it would not participate in peacekeeping operations and it would withdraw from diplomatic engagement in international conflicts, including the Arab-Israeli conflict.[33] Whether a state is a democracy or not would play no role in America's diplomatic relations with it.

Whichever of these two trends prevails, US foreign policy is likely to be marked by three defining features. First, the cult of pragmatism. America will remain the only power with a truly global outreach, its responsibilities will be broader than those of any other country and it will continue to be the primary target of belligerent actors intending to challenge the global order.

Reflecting its domestic conditions, America's strategic culture will remain driven by this cult of pragmatism, rather than principle. The primary preoccupation of any President of the US will be to remove any threat to US security. Whether this should be done by acting unilaterally or through the existing multilateral structures will be of secondary importance for the executive.

Second, America's attitude towards the use of force will remain on the whole permissive, notably in so far as the use of force for self-defence is concerned. The American definition of self-defence will remain expansive. On the other hand, the US will be increasingly reluctant to accept other kinds of deployments such as peace-keeping or post-conflict stabilisation, notably under the flag of the United Nations. Third, America's attitude towards international law will remain driven by its pragmatic considerations. In other words, it will be multilateralism *à la carte*. Unlike most Europeans, the United States continues to see domestic law as superior to international law. The reasons for this are again structural – as a superpower, the US sees international rules as constraining rather than empowering. If that is the case, however, a key question will concern whether and how the US will accept the further institutionalisation of international relations, in response to the challenges to (and potential decline of) its relative power in the face of emerging global actors.

Notes

1. It is noticeable, however, that three quarters of the increase in US government R&D spending since 2001 was allocated to the defence sector. See the *OECD Science, Technology and Industry Scoreboard 2005 – Towards a knowledge-based economy* (OECD, October 2005). Available online at: http://caliban.sourceoecd.org/vl=10771102/cl=17/nw=1/rpsv/scoreboard/.
2. *Academic Rankings of World Universities 2005*, published by the Institute of Higher Education, Shanghai Jiao Tong University.
3. 'Dynamic Sectors give global growth centres the edge', Deutsche Bank Research, October 2005.
4. John Williamson 'What Follows the USA as the World's Growth Engine?', India Policy Forum, India Policy Forum Public Lecture, 25 July 2005.
5. William R. Cline, *The Case for a New Plaza Agreement*, Policy Briefs in International Economics, Institute for International Economics, December 2005, and 'Economic survey of the US', OECD Policy Brief, 2005.
6. Kevin Phillips, *American Theocracy: The Peril and Politics of Radical Religion, Oil, and Borrowed Money in the 21st Century* (Viking, New York, 2006).
7. Andrew Kohut and Bruce Stokes, *America Against the World: How We are Different* (Times Books, New York, 2006).
8. See the report, 'Unauthorised Migrants: Numbers and Characteristics', Pew Hispanic Centre, 14 June 2005.
9. For example, 73.6 % of Mexican-born immigrants did not speak English well and 43% were 'linguistically isolated'. Linguistic difficulties are also evident among the second-generation immigrants. Samuel P. Huntington, 'The Hispanic Challenge', *Foreign Policy*, no. 141, March/April 2004.
10. Samuel P. Huntington, op. cit.
11. Kohut and Bruce, *America Against the World*, op. cit.
12. 'The Debate Over Immigration Reform', *The New York Times*, 9 June 2006. 'Bush's Speech on Immigration – Transcript', *The New York Times*, 15 May 2006.
13. Thomas Piketty & Emmanuel Saez, 'The evolution of top incomes: a historical and international perspective', *American Economic Review*, 2006, quoted in 'Special report: Inequality in America', *The Economist*, 17 June 2006.
14. Arthur B. Kennickell, 'A rolling tide: changes in the distribution of wealth in the US, 1989-2001', *Levy Institute Working Paper* 393, November 2003, quoted in Jacques Mistral & Bernard Salzman 'La préférence américaine pour l'inégalité', *En Temps Réel*, Cahier 25, 2006.
15. With the exception of Poland and to a lesser extent Italy and Spain – for comparative opinion polls see Kohut and Stokes, op. cit., p. 108.
16. Pew Research Center for the People and the Press, *Survey of Religion*, July 2005.
17. Pew Research Center, *Survey of Foreign Policy*, July 2004; Kohut and Stokes, op. cit., p. 115.
18. William Kristol and Robert Kagan, 'Towards a Neo-Reaganite Foreign Policy', *Foreign Policy*, vol. 75, no. 4, July/August 1996, pp. 18-32.
19. Kohut and Stokes, op. cit., p. 195.
20. Ibid., pp. 172-6.
21. 'Tony Snow: Polls aren't key to Iraq', *NewsMax.com*, 18 June 2006.
22. 'Bush offers Europe a defence of Iraq policy', *International Herald Tribune*, 22 June 2006.

23. Kohut and Stokes, op. cit., p. 30.
24. In the Pew Research 2005 poll of youth from 16 countries, only 10% choose the US as the 'land of opportunity'. See: Kohut and Stokes, op. cit., pp. 34-6.
25. Pew Research Center for the People and the Press, *Foreign Policy Attitudes*, July 2005.
26. Marcin Zaborowski, 'US China Policy: Implications for the EU', EUISS, Paris, October 2005. Pierre-Antoine Braud, 'La Chine en Afrique : Anatomie d'une nouvelle stratégie chinoise', EUISS, October 2005. Available at: http://www.iss.europa.eu/new/analysie.html.
27. Before coming to power, and then in the first year of this administration, President Bush argued in favour of a more assertive policy towards China and criticised the softer approach of his predecessor. Subsequently, however, no tangible policy shift occurred. For a typical 'China bashing' view, see Edward Timperlake and William C. Triplett III, *Red Dragon Rising: Communist China's Military Threat to America* (Regnery, Washington D.C., 2002).
28. Currently, 33% of oil imports come from Latin America, 23% from the Middle East, 18% from Africa and 16% from Canada. Concerning gas, US dependence is relative since 85% of gas imports come from Canada.
29. 'President Bush Delivers State of the Union Address', The White House, Office of the Press Secretary, 31 January 2006. Marcin Zaborowski, 'Softer Bush Exports Foreign Policy Dilemma to Europe', *European Voice*, 9-15 February 2006.
30. See the statement by General James L. Jones, USMC Commander, USEUR, before the House Armed Service Committee, on the relevance of Africa for US oil imports and the need to enhance US assistance to the African Union and other sub-regional organisations to preserve stability in the continent; quoted in Samantha L. Quigley, 'EUCOM Leader Calls Africa Global Strategic Imperative'; American Forces Press Service, Washington, 8 March 2006.
31. See the Report of the National Energy Policy Development Group in 2001, putting an emphasis on the need to enhance economic and political ties between the US and Africa with a view to promoting a more favourable environment for US energy trade and investment. It should be noted, however, that the biggest single recipient of US investment – South Africa – is not an oil-rich country, but a booming economy.
32. Some elements of this prospective foreign policy outlook are outlined in Francis Fukuyama, *After the Neocons: America at the Crossroads* (Yale University Press, 2006).
33. Philip H. Gordon, 'America's Role in the World: Searching for Balance', in Marcin Zaborowski (ed.), *Friends Again? EU-US relations after the crisis* (EUISS, Paris, 2006).

China

By any standards, China will be regarded as a pivotal player in the international system of the twenty-first century. China today is home to 20% of the world's population, generates almost 5% of its wealth, accounts for 12% of global primary energy demand, holds around $850 million in foreign reserves,[1] ranks as the world's second FDI destination,[2] and trades $1.4 trillion worth of goods and services.[3] Clearly, China is a major regional power with a global impact and a key player of global interdependence. By the 2025/2030 time horizon, according to most extrapolations of current trends, China's population will reach 1.44 billion and its energy consumption will skyrocket, with demand growing by 150% for oil, tripling for gas and almost doubling for coal. Massive energy imports and investment in infrastructure will be required to sustain economic expansion, which should lead China's GDP to triple by 2025 – second only to the US (except the cumulative GDP of the EU) and possibly overtaking it in PPP terms.

By then, China will have become the largest trading nation in the world: a commercial superpower exerting an unparalleled magnetic attraction throughout Eastern and Southern Asia – a region that will itself produce around 40% of global GDP in PPP terms in 2025. Already a key player in international relations, China may come to have a determining influence on the future global order, structuring global politics and economics in cooperation, or competition, with two or three other major 'poles' of power. China's transformation over the next twenty years will therefore entail far-reaching consequences for reshaping the international system.

The momentous implications of China's rise pose at least two critical sets of questions, addressing respectively the sustainability of the country's projected growth and China's priorities as a global player. The two are closely related, as the formidable economic, social and environmental challenges to development are likely to absorb China's elites for the foreseeable future and affect China's strategic outlook, putting a premium on stability. China's remarkable efforts to sustain economic expansion and improve its citizens' quality of life may, on the other hand, have disquieting implications at the international level, when it comes to weighing immediate economic imperatives against wider political, security and environmental considerations.

Domestic challenges

Energy security and power generation

Over the next twenty years, China will need to confront a range of formidable challenges to preserve economic growth and social stability, including securing energy supply and countering environmental degradation, reforming the welfare system and fighting corruption. The projected increase in energy demand will put strains on the supply system and on the country's generation capacity. China is aware of its vulnerable position, depending for almost 50% of its oil imports on unstable regions in the Middle East, and importing 80% of its energy needs via sea-lanes through the vital Strait of Malacca, the jugular vein of the emerging superpower. Against this background, China has begun to proactively pursue alternative supply options and routes, and will continue to do so as a matter of overriding national interest. China has struck important energy deals with four key countries, namely Saudi Arabia, Iran, Angola and Russia, which today supply 45% of China's oil needs , as opposed to only 13% in 1995.[5] In addition, Chinese interests in African oil resources is growing, with deals involving several countries such as Nigeria, Sudan and Libya. China is also cooperating in the development of port facilities in Pakistan and Myanmar, which will provide platforms for securing energy routes and for the projection of its naval power in the Indian Ocean.[6] On the Asian mainland, the development of a huge 'Asian power grid' is under active consideration, with proposals to extend the much discussed gas pipeline from Iran to India to reach China, and plans to tap into the vast Central Asian oil reserves through new pipelines via Uzbekistan and Kazakhstan.

Securing energy supply from abroad is, however, only one side of the problem. A potentially bigger challenge will consist in building sufficient capacity for power generation. To sustain its growth rates, it is estimated that China will need to quadruple its current electricity production, mounting from 360 GW to 1220 GW/year in 2030. Around two thirds of electricity will still be derived from coal (compared to around 80% today), of which China is one of the three biggest producers together with the US and India. Hydropower is expected to generate 200-240 GW, gas-fired plants 111 GW, renewables 38 GW and nuclear energy 35 GW. However, these projections need to be assessed in the light of both financial and environmental costs. Matching these estimates would entail an investment in energy infrastructure of $2 trillion over the next 20/25 years (12% of the world's total energy investment). Considering the many social and economic demands that China's budget will need to confront, this hardly seems feasible.

Environmental degradation and urbanisation

The growing focus on energy efficiency matches the pressing concerns with pollution and environmental degradation, which already costs China almost 10% of its GDP.[7] Due to its reliance on fossil fuels, China's share of world CO_2 emissions is expected to rise from 15.2% to 19% in 2025.[8] Combined with biomass burning, polluting emissions already cause acid rain on 30% of China's territory, a quarter of which is also exposed to rapid desertification.[9] Over the last 10 years or so, China has lost almost 8 million hectares of farmland[10] and the process is continuing at a pace of 200,000-300,000 hectares per year. The deterioration of the remaining arable soils affects one third of the total.[11] These developments have put strains on food supplies, with China importing 6% of its grain demand in 2000; however, projections for the future do not give rise to undue concern.[12]

Water shortages are certainly more serious, with 75% of major rivers polluted and 180 million people depending on contaminated water. Northern regions experience serious shortages, notably in the valleys of the Yellow, Huaihe and Haihe rivers. Water scarcity is such that official statements describe it as 'an unavoidable issue threatening national security'.[13] It has been estimated that environmental degradation will result in 20 to 30 million 'environmental refugees' up to 2020 (compared to 6 million today), adding to the huge number of economic migrants within the country.[14] Hence, social unrest linked to environmental degradation and water scarcity is likely to grow.

The proportions of the coming environmental crisis are such that massive intervention will be required to change energy production and consumption patterns, improve soil and water management and reform agriculture. Investments in water management alone have been evaluated at tens of billions of dollars and some huge projects, such as the north-south water diversion scheme,[15] have been launched. These policies will absorb considerable financial resources but, if seriously undertaken, will contribute to improving the quality of life of Chinese people – yet another key domestic priority in order to preserve social stability and cohesion. With a per capita income of about $1,700[16] and 300 million people lifted out of poverty over the last twenty years, China has accomplished considerable progress, but economic growth has exacted a heavy price in social terms.

At least 120 million people have migrated from the mainland to the East of the country and the burgeoning coastal regions, with income differentials between the poorest and the richest share of the population of an order of 1 to 8. The urbanisation rate is expected to grow from 41% to 57% in 2025. By then, the urban population will amount to over 800 million people, in a country where urban pollution has already reached alarming levels. The inadequate sanitary infrastructure will pose a serious threat with a considerable potential for spreading contagious diseases and growing air pollution, already responsible for 300,000 to 500,000 premature deaths per year. The ailing public health system is unlikely to be able to confront these challenges in the short term, with less than half of the urban population and only 10% of the rural population covered by public medical insurance.[17]

Economic growth and political power in China

China's impressive economic achievements tend to overshadow not only the environmental and social impact of breakneck growth, but also the inherent weaknesses of China's economic system. The state has a very strong grip on the economy, with state-owned, low-productivity enterprises estimated to account for almost 40% of the economy[18] and the Party co-opting new social elites to acquire their loyalty. This results in what has been defined as 'crony capitalism',[19] with political elites holding the reins of public enterprises operating on the market and reaping the benefits of economic power and notably of the privatisation process. The collusion of political and economic interests leads to the poor economic performance of state enterprises,[20] a weak banking system[21] and high levels of corruption.[22]

The key problem for the future is that social order depends primarily on the fulfilment of the ambitious goals of economic growth, and the consequent rise of average living standards, short of which the system might crack under the burden of its contradictions and dysfunctions. Resources will also need to be directed to setting up a credible pension system for an elderly population expanding from 7% today to 20% (or 300 million) in 2025, with a dependency ratio of 39 in 2015 and 50 in 2030. With a view to sustaining the pace of economic growth, innovation is regarded as key to accelerate the progress towards creating a knowledge-intensive society, but much ground needs to be covered to mobilise the necessary volume of investment in S&T and R&D and maximise output.[23] Innovation as such can surely help in the medium to long term, but not provide the main answer to the pressing social needs of a huge population, in a very diverse country. The ability of the political system to take account of the size of the problems at hand, address the weaknesses of economic governance and manage social tensions will be of pivotal importance.

China's impressive record of economic modernisation has not been matched by meaningful political reforms, and there is little indication that these may follow in the near future. China remains a one-party state, with pervasive state control over the media and a grim human rights record.

Looking back at the Tiananmen repression of 1989, some evidence of greater openness can be detected: some limited degree of criticism of the government is tolerated, as was the case during the outbreak of the SARS epidemic, and the rule of law is being reinforced, not least to fight corruption. However, these are minor concessions and their purpose is clearly to strengthen the legitimacy of communist rule rather than to move towards democratic transition.

The Chinese establishment justifies the authoritarian regime by pointing at the potential disorder ensuing from an excessively rapid opening up of the political system. But there is a risk that the mismatch between official policy guidelines and the situation on the ground is growing too large. In addition, the strong centralistic tendencies of the Chinese state stand in contrast to the multicultural and multi-ethnic reality of the country, including regions and nations as diverse as Tibet and Xinjing, whose respective populations have endured discrimination and forced assimilation.[24]

In parallel to the progressive extinction of Marxism as a driving ideology, renewed 'ideological hunger' has become an important factor of Chinese society and politics. The rise of (state-sponsored) nationalism is a worrying expression of this trend. On the other hand, a revival of organised religious movements can be detected, including the Falun Gong cult but also Islam, Christianity and Buddhism. The Communist Party does not object to these religions *per se* but to the fact that it does not control their organisations. Hence, the government is in the process of setting up 'alternative' state-controlled churches. This seems to be the pattern for the future – stepping back from the atheist anti-religious agenda, accepting (even supporting) the existence religious movements but imposing state control over the movements' organisations.

On the whole, although unlikely to evolve towards a fully-fledged democracy for the foreseeable future, the Chinese political system will probably become more participatory, more inclusive and more responsive to the major stakeholders, such as business groups, but also to some nongovernmental organisations, such as environmental groups. Paradoxically, the best chance of preserving the rule of the Communist Party, at least in the medium term, is a further injection of market economics to sustain economic growth and making resources available to address pressing environmental and social concerns. China is likely to evolve towards 'soft-authoritarianism' with one (no-longer communist in any meaningful way) party in the lead – a regime comparable to those in power in other countries such as Singapore. The success of China in sustaining its economic development and political stability will be decisive not only with a view to addressing its internal problems but also to boosting its influence in the world.

China and the world

The strategic outlook

China has always sought to appease international concerns stemming from its rise to power status, adopting different conciliatory formulas the latest of which describes China's current trajectory as a 'peaceful rise'.[25] Within the basket of principles that China vocally supports in international relations, 'non-interference' in the internal affairs of independent states stands out as a cornerstone of its vision of the global order. Whether one looks at the pursuit of the 'democratisation of international relations' (as opposed to Western-style internal democratisation), at the declared opposition to hegemonic designs and power politics, or at the discourse on a multipolar world, China's foreign policy narrative is implicitly defined by opposition to the proactive agenda of the only global superpower – the US.

It has been suggested that the original Chinese model of development, as opposed to the top-down package of ideas and policies commonly referred to as the 'Washington consensus', has attracted considerable sympathy in the developing world, as an instance of emancipation from Western influence.[26] Likewise, emphasis on multilateralism and on the unique legitimacy of the UN system plays to the advantage of China and of those other countries who regard multilateral institutions as a shelter against the perceived vagaries of US foreign policy. On the whole, Chinese foreign policy has been directed towards ensuring stability in its neighbourhood, and towards expanding China's scope for manoeuvre and capital of influence incrementally, through the peaceful means of skilful foreign economic policy. This consensus-oriented, non-judgemental foreign policy posture has won China a number of friends not only in the developing world, including the controversial regimes of Iran, Sudan, Zimbabwe and Venezuela, but also in Europe.

Three key questions

Looking at the future impact of China on international relations, the first question is whether China subscribes to this cooperative, multilateralist approach out of convenience or conviction. While the difference may be intangible in the short term (and any foreign policy contains a mix of both elements), it is highly significant in a long-term perspective. China supports loose multilateral frameworks, notably centred on economic cooperation, but is reluctant to commit to binding regimes, except when obvious advantages derive from them, as in the case of the WTO. China's permanent membership of the UNSC provides as much power as responsibility. So far, China has stopped well short of allowing legal or ethical considerations on human rights, human security or good domestic governance to affect its economic or energy interests. In pursuing its international energy policy, for example, the Chinese government and state energy companies show no regard for the political credentials of their interlocutors and often accompany energy deals with a range of economic, technological, cultural and also security policy measures, tempting the supplier countries into closer partnerships. This potentially undermines Western political conditionality imposed on unpalatable regimes.[27] Hence the related question, which could be asked looking at the security, energy, environmental, and commercial dimensions of external relations, whether China will accept the responsibilities that come with power – in other words, will it become a proactive 'stakeholder' as opposed to a powerful 'shareholder' of global governance?

The second question is whether, as a potential major stakeholder, China will endorse or distance itself from the global frameworks and agendas that the US and the EU have shaped, which include not only trade rules but also principles of democracy and freedom. In other words, the question is often asked whether China will be a *status quo* power or a revisionist actor. One may say that, due to its sheer size and cultural distinctiveness, China will simply change the *status quo*, willing or not.

The third and probably most important question concerns the direction of such change. China's identity as a fledging superpower is as yet blurred, but it can be argued that China will develop its own distinctive 'role', its own way. American, European and other foreign perspectives will be factored in in the calculation of national interests, but will in no way determine them. Perhaps of greater influence will be structural developments, including growing interdependence and the need to overcome a zero-sum approach to questions of collective public interest such as energy supply and environmental protection. China is conscious of the relatively weak domestic bases of its rising power and of the utility of a cooperative approach. The latter may take root and develop an institutional dimension beyond calculations based purely on political self-interest.

The regional dimension

Recent developments in the relationship between China and its neighbours, as well as the envisaged plans for further regional integration, offer a test case of the key guidelines, tensions and contradictions of China's foreign policy. On the one hand, China's foreign policy has been quite assertive at the regional level, notably concerning border issues with Vietnam, South Korea and India (a resolution of residual differences with India, the other Asian giant, seems, however, in sight). The question of Taiwan represents a factor of systemic instability, and the deterioration of Sino-Japanese relationships might lead to dangerous power politics in the region. On the other hand, however, China has actively sought to associate itself with existing regional bodies and conclude new broad-based cooperative agreements, as demonstrated by China's commitment to the ASEAN + 3 format and by the prospect of setting up the ASEAN-China Free Trade Area by 2010.[28] The East Asian Summit of December 2005, strongly sponsored by China, included all regional powers except the US.

This is a symptom of the Janus-faced nature of Chinese-led regional integration, directed towards promoting the common good but also towards setting US policy in a multilateral context that will make the pursuit of bilateral alliances more difficult and will ultimately reduce American influence. The consolidation of the Shanghai Cooperation Organisation seems to respond even more directly to the purpose of defining autonomous forms of regional governance, led in this case by China and Russia, in energy-rich Central Asia. A key engine of economic growth in neighbouring regions, China also supports outcast regimes in Myanmar/Burma and Uzbekistan, with little regard for the sustainable development requirements of its client states.

Notes

1. China overtook Japan as the first holder of official foreign exchange reserves in the first quarter of 2006.
2. In 2004, FDI stocks in China and Hong Kong amounted to almost $700 billion. See *World Investment Report: Transnational Corporations and the Internationalization of R&D* (UNCTAD, United Nations, 2005).
3. That corresponds to 7% of world trade.
4. Andrew Small, *Preventing the Next Cold War: A View from Beijing* (The Foreign Policy Centre, London, 2005).
5. Philip Andrews-Speed, 'China's energy policy and its contribution to international stability', in Marcin Zaborowski (ed.), *Facing China's Rise: Guidelines for an EU Strategy*, Chaillot Paper (EUISS, Paris, forthcoming).
6. The US Department of Energy provides a short overview of China's dynamism in the quest for new resources: 'In recent years, the Chinese have significantly increased the number and geographic distribution of energy assets and investments, although total overseas oil investments by Chinese firms remain small compared to investments by the international oil majors. 56 Chinese national oil companies have invested in oil ventures in over 20 countries with bids for oilfield development contracts, pipeline contracts, and refinery projects in Iran, Sudan, Kazakhstan, Kuwait and others. In addition, the Chinese have recently focused on broadening their equity stakes in North Africa, Central Asia, Southeast Asia, Latin America and most recently in North America, where they have acquired stakes in Canadian oil sands firms and unsuccessfully attempted to acquire the US firm Unocal.' Energy Policy Act 2005, Section 1837: National Security Review of International Energy Requirements, prepared by the US Department of Energy, February 2006.
7. The notion of 'Green' GDP is increasingly applied in the Chinese debate to subtract from 'Black' GDP the financial implications of environmental degradation, with a difference amounting to more than $100 billion. World Bank, 'World Development Indicators', quoted in 'L'energia al potere', *Aspenia*, no 32, 2006. See also Joshua Cooper Ramo, *The Beijing Consensus* (The Foreign Policy Centre, London, 2004).
8. Under a worst case scenario, short of implementing major policy programmes to promote efficiency in energy production and consumption, per capita annual emissions levels could reach 1.33 tons of carbon by 2020, which is more than double the per capita emissions of the year 2000 (but still much lower than OECD countries). *China National Energy Strategy and Policy to 2020, Subtitle 2: Scenario Analysis on Energy Demand*, Energy Research Institute, National Development and Reform Commission, China Sustainable Energy Program, May 2004 (http://www.efchina.org/home.cfm).
9. According to the Chinese Academy of Sciences (CAS), the total area of China's deserts is 1.57 million square kilometres. The northern arid lands of China are a subject of concern since a large part of them, according to official statements, are entering into the first stages of desertification.
10. Roughly equivalent to the area of Portugal.
11. Li ZhiDong, 'Energy and Environmental Problems behind China's High Economic Growth – A Comprehensive Study of Medium- and Long-term Problems, Measures and International Cooperation', Institute of Energy Economics (IEEJ), Japan, March 2003 and Chinese Ministry of Environment, *Report On the State of the Environment In China*, 2004 (http://www.zhb.gov.cn/english/SOE/soechina2004/land.htm).

12. *Poverty Alleviation and Food Security in Asia: Lessons and Challenges* (Food and Agriculture Organization of the United Nations, Regional Office for Asia and the Pacific, December 1998).
13. Nathan Nankivell, 'The National Security Implications of China's Emerging Water Crisis', *China Brief*, vol. 5, no. 17, The Jamestown Foundation, 2 August 2005.
14. See Norman Myers, 'Environmental refugees: an emergent security issue', 13th Economic Forum, Session III – Environment and Migration, Prague, 23-27 May 2005.
15. This project aims at linking large water basins to water-consuming regions. Three canals are to be constructed, from the Yangtze river to Western and Northern China. Costs are currently estimated at around $61.6 billion. Xinhua News Agency, 5 March 2003.
16. Or 15% of the US level in PPP.
17. C. Fred Bergsten, Bates Gill, Nicholas R. Lardy, Derek J. Mitchell, *China: The Balance Sheet. What the World Needs to Know Now about the Emerging Superpower* (Center for Strategic and International Studies and Institute for International Economics, Public Affairs, New York, 2006).
18. *Etude économique de la Chine, 2005* (OECD, Paris, 2005).
19. Minxin Pei, 'The Dark Side of China's Rise', *Foreign Policy*, March/April 2006.
20. Ibid. According to Pei, more than 35% of state enterprises lost money in 2003 and more than 1 in 6 had more debts than assets.
21. The restructuring of the banking sector is however already well underway. Recent FDI inflows are a good thing for the long-term prospects of the domestic banking sector. See Guonan Ma, 'Who Pays China's Bank Restructuring Bill?', *CEPII Working Paper*, 2006-04.
22. Ibid. According to Pei, in 2004, over 170,000 Party officials and members were involved in corruption scandals but only 3% of them were prosecuted, which would result in a huge degree of impunity and faltering rule of law. See Pei, op. cit.
23. Christopher J. Forster, *China's Secret Weapon? Science Policy and Global Power*, The Foreign Policy Centre, 2006, and Joshua Cooper Ramo, op. cit.
24. Ross Terrill, *The New Chinese Empire* (Basic Books, New York, 2003).
25. Zhen Bijian, '"Peacefully Rising"' to Great-Power Status', *Foreign Affairs*, vol. 84, no. 5, September/October 2005.
26. Cooper Ramo (op. cit.) has introduced the expression 'Beijing consensus' to mark the difference between the Chinese model of development and the traditional formulas prescribed by the IMF and the World Bank. According to the author, the Beijing consensus is based on three 'theorems', respectively addressing the value of innovation to boost development, the focus on quality of life and sustainability, and self-determination.
27. See Philip Andrews-Speed, op. cit.
28. China is currently taking part in two trade agreements in the region and negotiating five others.

India

The New Global Puzzle — PART II

11

India is a 'contrary' country not only because of its long-standing tradition of argumentation and debate, which are at the roots of its democracy,[1] but also because of the many facets of its uneven, yet impressive, socio-economic development. India is today the location of 25% of the world's offshored IT and IT-enabled services,[2] but India's economy is still heavily dependent on agriculture, which accounts for 22% of GDP and provides employment for 55% of the labour force.[3] Although the Indian middle classes (150-200 million people)[4] are eager to fully participate in the processes of global production and consumption, one third of India's population (360 million) still lives in extreme poverty on less than $1 per day.[5] In a country with an adult literacy rate of only 70% for men and less than 50% for women, the academic system, which comprises 380 universities and 1,500 research institutes, generates an annual output of 300,000 engineers and science graduates, and around 10,000 Ph.Ds.

These and many other contradictions should not come as a surprise when looking at a diverse federation of 28 states (and 7 Union Territories) which are heterogeneous in terms of language, wealth, human development, political dynamics and ethnic and religious identities. What is most remarkable is that this huge country of over 1 billion people has managed to preserve and consolidate a functioning democratic regime and, over the last 15 years, to open up its economic system and more than double its GDP. Growth is projected to continue at a fast pace, between 6% and 8% on average, for the next two decades. These growth rates would make India in 2025 the fourth largest national economy in the world, behind the US, China and Japan.

Domestic challenges

Political and social stability

The sustainability of growth will, however, be put to serious test over the next 20 years, with the Indian population growing by around 350 million, energy dependency on foreign supplies reaching nearly 50% for gas, and over

90% for oil,[6] and public spending required to double, triple or increase even more dramatically in critical sectors such as transport, health and education among others. The question is, essentially, whether India will be able to convert its growing national income into a benefit for the whole population, and lay the foundations for becoming a stable middle-income country over the long term. From a political standpoint, the progressive shaping of a more inclusive and less centralised system of governance bodes well for the future of Indian democracy. Following decades of secessionist tensions, the devolution of economic and fiscal authority, as well as more political autonomy, to regional governments has defused conflicts and paved the way for responsible regionalism.[7] In a society where castes still represent an important aspect of individual identity, a 'silent revolution' has taken place, with traditionally marginal classes and tribes[8] outstripping the higher classes in terms of voter turnout at elections and converging to support new political formations, which have seized power in some of the largest states of the federation. The 'politicisation of the castes' has led to the extension of positive discrimination and of the system of quotas in favour of the lower classes, leading to a new wave of 'democratisation' of India's democracy.[9]

Renewed political stability, following the Hindu-Muslim clashes in Gujarat in 2002, seems to have been established on solid foundations, although two problems loom potentially threatening the cohesion of Indian society in the years to come. On the one hand, disparities between and within regions are widening, with investment concentrating around a few large urban centres.[10] Internal migration is expected to grow, putting strains on overcrowded conurbations and potentially leading to conflicts between different ethnic groups. On the other hand, the Muslim minority of around 130 million people (12% of the population) remains at the margins of Indian politics and society. Failure to raise the levels of education and employment among Muslims, and to encourage their political representation, may offer fertile ground for political and religious extremism. On the whole, it appears that the next big challenge for India will be social and economic reform and cohesion, as a condition for sustained growth in the long term. In this perspective, the 'India system' will be put under stress by demographic expansion, the shortcomings of the healthcare and education systems, urban overcrowding, inadequate infrastructure and energy dependency. These challenges will notably require a thorough overhaul of India's faltering administrative framework, which is increasingly falling short of providing essential public services such as education and health.[11]

Demographics, growth and investment

With high fertility rates and longer life expectancy (70 years in 2025), the Indian population will almost match the Chinese population by 2025 and probably bypass it in the following decade. India's advantage compared to both China and the developed world will reside in its young population, with a projected median age of only 30.4 years in 2025, and less than 34 in 2035. The old-age dependency ratio, with an elderly population (over 65) doubling to about 80 million by 2025 (only 6% of the population) will remain very low as well. Endowed with the largest pool of young population in the world (half of India's population is under 25), India will account, over the next five years, for almost one quarter of the global increase in the working-age population.[12] The expansion of the Indian workforce will accelerate from 7 to 15 million additional workers per year, which means that 200-250 million jobs would be required to absorb labour supply in the next two decades.[13] The problem is that the most dynamic economic sectors are also the least work-intensive: out of a current workforce of 400 million people, only 3 million are employed in the IT sector. That said, other important innovation-intensive economic sectors, like biotechnology and pharmaceuticals, offer promising prospects, with India already featuring among the biggest global producers of generic drugs. In addition, because of its young, expanding and cheap workforce, India is becoming increasingly attractive for investment in the manufacturing sector as well, potentially challenging the position of China as 'workshop of the world'.[14] India's multinationals are also increasingly active abroad, as the recent takeover of Arcelor by Mittal Steel, thereby creating the world's largest steelmaker, testifies. At this stage, however, it is as yet unclear whether India will be able to foster the expansion of a sizeable industrial domestic power base. Industry only accounts for 27% of India's GDP, with services accounting for 50%. Economic governance reforms, including loosening stringent labour regulations and dismantling the privileges of some industrial sectors, are considered to be among the key steps necessary to trigger the Indian industrial revolution.[15] The fulfilment of India's huge economic potential crucially depends, however, on considerable investment in its human and physical resources, and on its energy policy.

Only 60% of young Indians are enrolled in secondary education (with a backlog of 300 million illiterate adults), and just 5% of the country's young workforce has undergone vocational training. The high number of graduates and engineers needs to be put in perspective too, since it is estimated that only 10 to 20% of them can compete on the international market. The share of scientists and engineers as a percentage of population is three times larger in China than it is in India.[16] For knowledge and highly skilled human resources to remain a factor of competitive advantage in innovation-intensive industries, India will need to massively invest in education (the doubling of public expenditure is foreseen) and in R&D (where India's investment is only 1/60th of that of South Korea).[17] Other basic factors of human development will need to be addressed in order to create a favourable socio-economic (and consequently investment) environment, first and foremost the availability of adequate healthcare. Only 35% of the population has access to essential drugs and medicines, with considerable disparities between rich and poor, as well as between cities and rural areas. Communicable diseases represent a serious threat to development, with 2 million new cases of malaria reported annually and the AIDS-related death-toll possibly rising to 12.3 million between 2000 and 2015.[18] It is envisaged that health expenditure may be quadrupled over the next two decades in order to confront these serious challenges.

Critical investment is also foreseen to upgrade the country's poor transport infrastructure, which risks putting brakes on the potential for economic growth. Estimated costs range between $150 and 200 billion over the next 10 to 15 years,[19] with a view to coping with freight traffic expected to grow fivefold between 2000 and 2020 and with passenger traffic due to quadruple. The strain on infrastructures will be particularly severe in urban areas, whose population will grow to around 550 million, or 40% of the country's total, by 2025. The problem is compounded by the fact that growth will notably impact on 60 to 70 large urban centres of over one million people, in a country where one quarter of the urban population does not have access to sanitation facilities and public/private investment is highly concentrated around key economic hubs.[20] At the same time, 56% of Indian households are not connected to electricity, and power cuts are recurrent.[21]

Growing dependency on energy supply

Of the various factors putting the 'India system' under stress over the years to come, energy dependency will be among the most critical. Energy demand is expected to increase by 109% in 2025, and dependency on imported oil and gas will drastically increase. Oil accounts for one third of India's total energy consumption. Domestic oil production, which covered half of the consumption in the 1980s, will only satisfy 10% of the demand ten years from now.[22] About 62% of Indian oil imports come from four countries: Saudi

Arabia, Kuwait, Iran and Nigeria. Gas consumption is expected to explode, growing by 292%. Discovery of new reserves, however, could allow India to cover nearly half of its demand. Specific political and geographical constraints are pushing India to invest in LNG, with several projects which by 2030 could cover 27% of the national demand. Most of the imports will probably originate from the Middle East (notably from Qatar, Oman and Yemen, but also from Australia and Malaysia). Coal should nevertheless remain the main primary source of energy, particularly in power generation (64% of electricity generation in 2030). Despite the important investments needed to upgrade the coal plants, rising oil and gas prices should maintain India's reliance on coal, since this country owns 10% of the world's total reserves.

Electricity output is already insufficient to satisfy the average energy demand. India's electric power demand is expected to more than triple by 2025, which will require the tripling of the installed generation capacity from 124 GW in 2005[23] to about 300 GW. With a view to the growth in demand, major investment is planned in the renewable and nuclear sectors.[24] The required annual investment in new capacity for power generation is estimated to amount to 2% of GDP up to 2030.[25] At the very least, two thirds of power will come from fossil fuel sources (coal, oil and gas).

India is in the process of deploying a multi-pronged strategy to address its vulnerability with regard to energy, including four main elements. First, India envisages intensifying the development of the nuclear share in power generation: eight new reactors will be added to the 15 existing ones. Ultimately, by 2050, the nuclear sector is expected to produce roughly 20% to 25% of the country's power generation. However, given the lack of national reserves of uranium, India will be confronted with a dilemma between relying on the international market to buy uranium, or investing in an alternative option, based on fuel breeding and reprocessing and, in the medium term, on the thorium fuel cycle. The pros and cons of this choice need to be assessed against the terms of the US-India nuclear deal, finalised in March 2006 and illustrated below.

Second, some envisage domestic alternatives to imported fossil fuels, namely the massive development of energy plantations to draw energy from biomass, as well as sustained investment in the production of biofuels, following the example of Brazil. Third, India is pursuing a more proactive policy of exploration of new oil and gas fields in its own territory. Fourth, India is concluding bilateral deals with neighbouring nations with a view to securing supply for the long term. Aside from the envisaged pipeline connecting Iran to India via Pakistan (a project severely opposed by the US), India signed a massive gas deal with Iran in 2005 to begin LNG imports in 2007, for a period of 25 years.[26] Gas imports will also flow from Burma, where Indian firms have substantially invested in newly discovered reserves, and probably from Bangladesh, with proven gas reserves of 14 trillion cubic feet.

Foreign policy

Energy security-driven foreign policy reflects only one dimension of India's growing international dynamism. A rising power with the potential and the vocation to make a major contribution to global governance, India is seeking to define its place in international relations. It has been stressed that at least two of the three factors that constrained India's foreign policy and global role since independence – the breakaway and enduring hostility of Pakistan, its own domestic socialist system, and India's closeness to Russia during the Cold War – have gone. This has determined a corresponding expansion of India's scope for manoeuvre.[27] Since the end of the Cold War and the opening of its economy to globalisation, India's foreign policy discourse has been shifting from the traditional non-aligned approach to a more pragmatic one. The latter applies to relations with major global players, with turbulent neighbours and with regional organisations across Asia. India shows a rhetorical allegiance to multilateralism but, in practice, has proved at best sceptical towards some of the flagship EU-led initiatives aimed at, for example, regulating the use of force, including the International Criminal Court and the Ottawa Convention to Ban the Use of Antipersonnel Landmines.[28]

India and the US: a fledgling strategic partnership?

India aspires towards Great Power status and seems attracted by the attributes of power, defined in the classic terms of economic and military means, among others. India by and large recognises that the US will remain the most powerful nation in the world for the foreseeable future, and seems comfortable with reaping the benefits of unipolarity while preserving and widening the scope for autonomous manoeuvre.[29] According to a 2005 Pew Global Attitudes Poll, the share of Indians expressing a positive attitude towards the US had risen from 54% in 2002 to 71% in 2005, a marked countertrend with respect to most other countries in the poll.[30] In its relationship with the US, India's political capital has begun to yield considerable returns (including the 'democratic dividend' stemming from the stability of its democratic regime).[31] The India-US agreement on 'Next Steps in

Strategic Partnership' of 2004 includes cooperation on space programmes, dual-use technologies and also civilian nuclear energy. Following the June 2005 strategic military cooperation agreement with the US, the US-India nuclear deal may offer India the opportunity to escape the tight constraints imposed by its refusal to subscribe to the NPT and the CTBT.[32]

According to the agreement, the Nuclear Supplier Group embargo on India will be lifted, which would notably entail India's ability to import uranium and broaden its options for the development of nuclear power, whereas India would bring all its civilian nuclear installations under IAEA safeguards and support the conclusion of the envisaged Cut Off Treaty.[33] Its military assets, however, would be exempted from external control. This deal could have far-reaching consequences. It will certainly open a huge nuclear market to the deadlocked western nuclear industries and could make the proliferating thorium fuel cycle option, based on fast breeder reactors, less interesting for India. But it will equally enhance Indian military capacity, raising Pakistani concerns and undermining China's current strategic nuclear superiority. Most of all, the deal sends ambivalent signals concerning the future of the NPT regime since it proves that the indigenous development of nuclear weapons is a credible long-term option. The claims of the US and others that India is not a horizontal proliferator,[34] advanced to justify the exception, could be counterproductive and prompt some states to quit the NPT and to develop their own capacity without proliferating towards other countries.

The deal, however, needs to be set against the big picture of the emerging India-US strategic relationship.[35] India's friendship will be of strategic relevance to the US to balance China's rise and to provide a bulwark against the spread of Islamic fundamentalism in South Asia.[36] India's military modernisation is also proceeding at an accelerating pace, with an expanding submarine fleet, the acquisition of a Russian aircraft carrier, a new ambitious ballistic missile programme, armament deals with Israel, and the envisaged acquisition of F-16 and F-18 fighter aircraft from the US.[37] The American courting of India has just begun, and much more is likely to be on offer to ensure that India swings towards the US in the Asian strategic balance.

The regional context

While the strategic relationship with the US seems set on very solid grounds, however, India is likely to selectively engage with different partners depending on the issues at hand. India perceives the relationship with the US not only as an insurance policy in case things turn sour with long-time rivals

Pakistan and China, but also and above all as a key leverage in bilateral negotiations with them, as well as a major asset in establishing its power status. The dialogue with Pakistan has taken on new vigour since 2004, following the last eruption of violence in Kashmir. Confidence-building measures have been multiplied to enhance economic, political and cultural exchanges, with a window of opportunity on both sides to enter serious talks on the definitive resolution of the border dispute, under considerable pressure from the US. Confronted with the threat of Islamic terror and with the triple challenge of demographic explosion, economic development and energy security, the two countries may consider that they can no longer afford a state of latent conflict.[38]

At a regional level, India has taken a proactive stance to extend its influence. Having set the stage for confidence building with Pakistan to the West, India is increasingly turning to the East. The 'Tsunami diplomacy' performed in the aftermath of the natural catastrophe in the Indian Ocean, with India providing financial help to other affected countries and dispatching its navy to deliver relief to Indonesia, sent an important political signal.[39] At the institutional level, the 2004 India-ASEAN partnership for peace, progress and prosperity is due to evolve into a free trade area including Brunei, India, Indonesia, Malaysia, Singapore and Thailand as of 2007, and expand to Burma/Myanmar, Cambodia, Laos, Vietnam and the Philippines in 2016. Closer economic links have also been established in 2005 with Singapore, the most important commercial partner in the sub-region, and 2006 will see the launch of the South Asia Free Trade Agreement (SAFTA), an emanation of the South Asian Association for Regional Cooperation (SAARC). The Bangladesh, India, Myanmar, Sri Lanka, Thailand Economic Cooperation organisation (BIMST-EC) provides yet another layer in the deepening framework of regional governance. Institutional consolidation both reflects and stimulates the growth of inter-regional trade flows. Since 2004, ASEAN +3 (China, Japan, South Korea) nations have become, as a group, the biggest commercial partner of India, accounting for 20% of its trade. The EU follows closely at 19%, with the US remaining the single biggest national partner with 11% of total trade in 2004/2005 and China featuring second, following a stunning 80% increase in the volume of trade between the two countries over the same year.

India-China: from tactical rapprochement to long-term partnership?

Closer economic ties and the new political dialogue with China disclose potential for a future partnership beyond the ongoing tactical manoeuvring, which is directed at improving economic relations but leaves some open political issues unaddressed. The future of the Sino-Indian relationship is hardly foreseeable, but one should distinguish short-term obstacles from long-term trends. Presently, China and India do not trust each other enough to commit to a structured strategic partnership. Steps to consolidate their positions in the Indian Ocean to secure vital commercial routes and energy provisions, for example, seem to reflect more competition than cooperation. The same could be argued looking at the Chinese and Indian commercial, economic and cultural penetration in South East Asia. The weight of the 'US factor' in their respective calculations, furthermore, is still preponderant, whether it is perceived as an opportunity or as a potential threat (or a mix of both). Much will depend on the ability of the US to successfully pursue its strategy of *divide et impera* on the Asian stage, based on bilateral deals essentially aimed at the containment of China. In addition, since 2005, India and Japan are laying the foundations of closer political, economic and also military relations – a move justified at least in part by their intent to counterbalance China's weight in the region.[40]

On the other hand, deeper structural forces can be detected, potentially leading to a convergence of the strategic outlook of the two rising powers over the medium to long term. A new strategic partnership for peace and prosperity was heralded in some quarters on the occasion of the visit of the Chinese Prime Minister to India in 2005. While that might be wishful thinking for the time being, the Sino-Indian relationship is growing fast from an economic standpoint. As mentioned above, China is the second largest commercial partner of India after the US and, if present trends continue, it will very soon become commercial partner number one. Joint energy exploitation projects are ongoing in Sudan and Iran, and India has obtained observer status in the Shanghai Cooperation Organisation at a time when options for Sino-Indian joint ventures in Central Asia are under review.[41] Deepening inter-regional integration, based on growing cooperation and economic complementarity, might contain the seeds of an enhanced partnership. China and India have a common underlying purpose to seek emancipation from an international system shaped by foreign powers, and share a claim to pursuing original development models, very much influenced by their distinctive cultural traditions. While, therefore, it is surely premature to speak of the strategic convergence of the two Asian giants, this should be regarded as a distinct possibility in the long-term perspective.

Notes

1. Amartya Sen, 'Contrary India', in *The Economist: The World in 2006*.
2. *World Investment Report 2004* (International Monetary Fund). India is second only to Ireland in this regard. If IT-enabled services only are considered, its world market share is even stronger (67%).
3. Sophie Chauvin & Françoise Lemoine, *L'économie indienne: changements structurels et perspectives à long terme*, CEPII Working Paper no. 2005-04, April 2005. The share of workforce in the agricultural sector is expected to drop to 45% by 2020.
4. Estimates of the size of India's middle class diverge widely. Taking an annual income of $2,000 or more as a reference, at least 150 million people can be included in the middle class. Defining, on the other hand, the middle-class as households having an annual income between $4,000 and $21,000, the National Council of Applied Economic Research (NCAER) found that the middle class included only 10.7 million people in 2001-2002, and expected it to increase to 28.4 million in 2009-10. See http://www.ncaer.org/.
5. *World Development Report 2006: Equity and Development* (The World Bank & Oxford University Press, Washington D.C. & Oxford, 2005).
6. But India will remain self-sufficient for coal, which will still be the main primary energy source in 2030.
7. Interview with Sunil Khilnani, 'La société indienne: tensions et transformations', *Questions internationales*, no. 15, La Documentation française, Paris, septembre/octobre 2005.
8. These include the 'other backward classes' (low castes), 'scheduled castes' (formerly known as the 'untouchables') and 'scheduled tribes'.
9. Max-Jean Zins, 'La plus grande démocratie du monde?', *Questions internationales*, no. 15, La Documentation française, Paris, September/October 2005.
10. It should be noted, however, that according to the Gini index of income inequality, India is a more equal society than other emerging economies, and developed countries. In particular, India scores 33, while the US scores 41, China 45 and Brazil as much as 59. See Gurcharan Das, 'The India Model', *Foreign Affairs*, vol. 85, no. 4, July/August 2006.
11. Gurcharan Das, op. cit.
12. 'Can India Fly?', Special Report, *The Economist*, 1 June 2006.
13. According to a more conservative estimate, India's labour force of 375 million in 2002 is set to expand by 7 to 8.5 million per year, with an overall increase of 160-170 million workers by 2020. See Report of the Committee on India Vision 2020 (Planning Commission, Government of India, New Delhi, 2002).
14. Christophe Jaffrelot, 'L'Inde, la puissance pour quoi faire?', *Politique internationale*, October 2006.
15. Gurcharan Das, op. cit.
16. By the same standard, India lags at 1/100th of US levels. Report of the Committee on India Vision 2020, op. cit.
17. For example, India's investment in nanotechnology is projected to grow from $2 billion in 2002 to $10 billion by 2010. Other R&D priorities include genomics, bioinformatics, DNA technologies, clinical studies and genetically-modified crops. See Report of the Committee on India Vision 2020, op. cit.

18. Worst-case scenarios envisage as many as 50 million casualties from AIDS between 2015 and 2050. Nevertheless, a recent study published in *The Lancet* journal shows a regional decrease in the prevalence of HIV infection for several years, putting worst-case scenarios in perspective. 'Trends in HIV-1 in young adults in south India from 2000 to 2004: a prevalence study', *The Lancet*, vol. 367, Issue 9517, 8 April 2006. On the dynamics and impact of HIV infection in India, see also *HIV/AIDS. Country Profile: India* (UNAIDS, Population Division of the Department of Economic and Social Affairs of the United Nations Secretariat, March 2003); *World Population Prospects: The 2002 Revision. Highlights* (United Nations, New York, February 2002); *The next wave of HIV/AIDS: Nigeria, Ethiopia, Russia, India and China* (The National Intelligence Council, September 2002).

19. Given the already high budget deficit, private investments, and particularly FDI, will be needed to finance big infrastructure projects. The recent assignment of Delhi and Mumbai airports to private consortia seems to bear out the trend. According to a report by Morgan Stanley, the overall investment required to upgrade electricity, roads, airports, seaports and telecommunications is much larger. India is likely to gradually increase its spending to $150 billion per year by 2010 (i.e. an average of 8% of GDP, from an initial level of 3.5% of GDP in 2004). See Chetan Ahya and Mihir Sheth, *India. Infrastructure: Changing Gears*, Morgan Stanley Report, 2005.

20. Report of the Committee on India Vision 2020, op. cit.

21. 'Can India Fly?', *The Economist*, 2006, op. cit.

22. In 2003 India was already depending on oil imports by 70% (IEA, *World Energy Outlook 2004*). Brahma Chellaney, 'India's Future Security Challenge: Energy Security', in *India as a New Global Leader* (The Foreign Policy Centre, 2005).

23. Data collected from the website of the Ministry of Power, India, accessible at http://powermin.nic.in/JSP_SERVLETS/internal.jsp.

24. The current power generation mix looks as follows: gas 12.6 GW (10%), coal 68.5 GW (55%), oil 1.2 GW (1%), hydro 32.3 GW (26%), nuclear 3.9 GW (3%), renewables 6.1 GW (5%).

25. *World Energy Outlook 2004* (International Energy Agency, Paris, 2004).

26. The mushrooming of new LNG import terminals on the West Coast of India shows the importance of envisaged energy supply from Iran and Oman. See Chellaney, op. cit.

27. C. Raja Mohan, 'India and the Balance of Power', *Foreign Affairs*, vol. 85, no. 4, July/August 2006.

28. Christophe Jaffrelot, 'Inde: un tropisme américain aux dépens de l'Europe ?', *Le Monde diplomatique*, September 2005.

29. Sunil Khilnani, 'India as a Bridging Power', in *India as a New Global Leader*, op. cit. and Gilles Boquérat, 'Une puissance en quête de reconnaissance', *Questions internationales*, no. 15, La Documentation française, Paris, September/October 2005.

30. C. Raja Mohan, op. cit. It should be stressed, however, that in 2006 the Indian perception of the US deteriorated, with the share of those having a favourable opinion of the US falling to 56%. See the 2002 *Global Attitudes Survey* (Pew Research Center for the People and the Press, April 2002) and *15-Nation Pew Global Attitudes Survey* (Pew Research Center for the People and the Press, 2006).

31. Sunil Khilnani, 'India as a Bridging Power', op. cit.

32. At the time of writing, however, the deal is under scrutiny at the US Congress. The Congress approval of Resolutions S. 2435 (Energy Diplomacy and Security Act), as well as S. 2429 and H.R 4974, which waive some legal restrictions on US nuclear transfers to India, is a prerequisite for the application of the agreement.

33. The Cut Off Treaty, under negotiation, prescribes a ban on the production of weapons-grade fissile material.
34. Horizontal proliferators are those that disseminate nuclear weapons or nuclear technology for military applications to other countries. Vertical proliferators are those who develop nuclear weapons programmes within their own country.
35. For an American perspective on the relations between India and the US, and the role of the nuclear deal therein, see Ashton Carter, 'America's New Strategic Partner?', *Foreign Affairs*, vol. 85, no. 4, July/August 2006.
36. See 'Joint Statement Between President George W. Bush and Prime Minister Manmohan Singh' on the occasion of the visit of the latter to the US, 18 July 2005, and the exchange between the two leaders in 'President, Prime Minister Singh Discuss Growing Strategic Partnership', on the occasion of the visit of President Bush to India in March 2006. Both these documents, and many others concerning the strengthening partnership between the two countries, are available at http://www.whitehouse.gov/infocus/globaldiplomacy/.
37. Christophe Jaffrelot, 'L'Inde, la puissance pour quoi faire?', op. cit.
38. Whether the momentum for peace will prove sustainable will depend, however, on domestic developments in Pakistan. Pakistani society remains pervaded by hostility towards India (and powerful forces exist in India against reconciliation) and the influence of Islamic fundamentalism is on the rise, putting the sustainability of the current pro-American, stability-oriented military regime into question. See Mariam Abou Zahab, 'Pakistan : entre l'implosion et l'éclatement ?', *Politique étrangère*, 2, IFRI, Paris 2006. See also 'Too much for one man', A Survey of Pakistan, *The Economist*, 8 July 2006.
39. Laurent Gayer, 'Conflits et coopérations régionales en Asie du Sud', *Questions internationales*, no. 15, La Documentation française, Paris, September/October 2005.
40. C. Raja Mohan, op. cit.
41. Laurent Gayer, op. cit.

Latin America

How Latin America evolves between now and 2025 will depend on three main factors: internal political developments, its relationship with the United States, and its ability to integrate further in the globalisation process. Latin America's population will not experience any drastic increase between today and 2025. While Asia's population will represent about 60% of the world population by 2020, and Africa will contain 17% approximately, Latin America and the Caribbean will represent 8.5% according to median estimations.[1] Relative demographic stability in Latin America may be an enabling condition for political stability and economic growth. Moreover, the fact that Latin America is geographically remote from the major hot spots of strategic conflict may be a blessing for the future. Despite this relatively positive context, however, Latin America will be confronted with considerable economic and political challenges.

Regional issues

The economy

The Latin American region has two major population centres, Brazil (188 million inhabitants today, set to rise to 230 million by 2025) and Mexico (107 million today, set to rise to 130 million in twenty years' time). Their foreseeable evolution points in different directions. By 2025, Mexico will be increasingly attached to the North American economic pole, given the volume of trade between those two countries and the sustained migrant inflow from Mexico to the United States, whereas Brazil will probably evolve into an economic and political pole on its own and foster relationships with its neighbours. Other focal points will be the Andean region, endowed with substantial energy resources, and the southern tip of the continent, where the economic growth of Chile seems set on solid foundations.

High growth rates have been recorded in the region in 2004 and 2005 (6% and 4.4% respectively), largely sustained by high commodity prices. This good performance, which occurred in a favourable international context, is however too recent to suggest a sustainable acceleration of the growth potential of the region (average growth rates stood at 1.1% in the 1980s and 3.3% in the 1990s). According to current estimates, growth should average 3% until 2020, well below the performance of emerging Asian countries.[2] Institutional as well as physical infrastructures need to be strengthened. In comparison to other developing regions, little progress has been achieved in upgrading infrastructure over the last decade. Endowed with a more extensive framework of productive infrastructures (road, electricity and telecommunications) in the 1980s, Latin American countries have lost ground compared to fast growing economies such as South Korea and even China, which is by far poorer in terms of per capita income. Expenditure on infrastructures currently averages less than 2% of GDP in the region, against an estimated 4-6% required over the next 20 years to catch up with the level of South Korea. These investments could boost growth and help to reduce inequality,[3] but the strong engagement of public authorities will be decisive, not least to encourage the flow of private investments.[4]

Latin America nevertheless remains a major FDI destination, attracting 26% of FDI directed to developing countries in 2005, for an amount of $61 billion. Mexico, Brazil and Chile rank among the top ten recipient developing countries.[5] FDI in the region increasingly targets the manufacturing sector, but remains largely directed to the natural resource extraction industry in a number of countries. The region holds important mineral and biological reserves and the agricultural as well as the mining sectors are well developed and internationally competitive. Nonetheless, the degree of economic diversification remains low and, with the exception of Mexico, all countries of the area (Brazil included) are specialised in commodities. International demand, and more specifically China's demand for ferrous and non-ferrous materials, could increase dependency on the export of commodities.

Energy

Latin America produces 9% of global energy. More specifically, the region produces 15% of world oil (and exports almost 50% of it) and 7.5% of world gas. Endowed with almost 9% of the world's oil reserves, and 4.5% of gas reserves, Latin America will play a key role in the future politics of energy supply to developed and emerging countries.[6] Considerable differences exist, however, between the countries of the region. Only Venezuela has large shares of oil and gas resources. Outside the Middle East, which contains about 61% of oil reserves, Venezuela is the largest world reservoir, with 6.6% of proven oil reserves, ahead of the Russian Federation (6.2%). Brazil also possesses a significant 1% of the world's oil reserves and Ecuador and Argentina, 0.4 and 0.2% respectively. As far as natural gas is concerned, the Middle East (40%) and Russia (26.6%) top the list, while Venezuela has 2.5% of the world's gas reserves, and other Latin American states possess some residual, yet important, shares: Bolivia (0.9%), Argentina and Trinidad & Tobago (0.3% each), and Brazil and Peru (0.2% each). From an energy standpoint, the region is highly integrated: different countries have either achieved self-sufficiency (such as Venezuela, Mexico, and Bolivia) or depend only on regional resources.[7]

Competition for Latin American energy resources is likely to grow, with the US envisaging to increase its imports from the region, and China expanding its presence, energy investments and joint ventures with local companies in countries such as Venezuela, Peru, Argentina, Ecuador and Brazil. Brazil, for its part, is emerging as a major regional actor in the energy sector. The main Brazilian oil company, Petrobras, is deeply involved in the exploitation and distribution of resources in Bolivia and Argentina. The relative economic weight of Brazil will prove a key factor in shaping the South American energy market. The 'Energy Ring' project, a $2.5 billion gas pipeline, which could connect the Peruvian and Argentinian gas fields to Brazil, Chile and Uruguay around 2010, illustrates this trend. Brazil is equally the primary world producer of ethanol and would become one of the main exporters of this biofuel, should it become more extensively used as a substitute for oil in developed economies.

Paradoxically, high oil prices coincide with a significant twist in the energy policy of some countries in the region, notably those where international oil companies control the extraction of resources (specifically Venezuela, Bolivia and Ecuador). In Venezuela, the re-nationalisation of resources may have set in motion a new political trend.[8] Bolivia and, to a lesser extent, Ecuador have followed this path. If sustained over the long term, these policies may pose a serious obstacle to the economic development of these countries. Investment in the energy sector will be greatly needed in order to meet the expected demand of importing countries (notably the US and Brazil – for gas – and, to a lesser extent, China). The overall investment required in energy infrastructure is quantified at $1 trillion over the next twenty years. Liberalising this sector is considered an essential precondition for attracting international investment and boosting national production capacities.

Political developments

The Cold War was a grim episode for Latin America. Marxist revolutions resulted in both repressive regimes and American interventionism. In the 1990s the situation changed, as transitions to democracy succeeded in key countries, allowing for greater freedom and respect for human rights. At the beginning of the twenty-first century, successive elections in Argentina, Brazil, Chile, Colombia, Mexico, Peru and elsewhere have consolidated mature democratic systems. Nevertheless the question remains as to whether these transitions have paved the way towards the permanent and irreversible democratisation of Latin America or whether setbacks to this process may still occur.

The two main challenges to Latin American democracy in the future will be the fight against inequality and the fight against political violence, organised crime and terrorism. In Latin America, the gap between the richer and poorer sections of society is very pronounced. On average, 40 to 47% of the total income is received by 10% of the population whereas only 1 to 2% goes to the bottom 20% of the population. Poverty is widespread in Latin American societies. The share of poor people in the region averages 24.6% of the population (30% to 31% in Mexico, Central America and the Andean Community, and 19% in the southern part of the continent).[9] It can be expected, however, that democratically accountable governments will, over the medium term, contribute to balance those economic disparities and expand the middle class. The problem is that populist political forces across the region promise to accelerate that process and introduce equality overnight through questionable policies. The poorer classes of society have showed an inclination to endorse those promises in democratic elections. But those policies may have ruinous consequences for the countries involved. The current economic policies of Venezuela and Bolivia, as noted above, jeopardise foreign investment and their countries' future development. It remains to be seen whether these two cases are just exceptions or whether they are setting precedents for a larger trend. Be that as it may, a reasonable forecast for the region as a whole is that the progressive consolidation of democratic practices will reduce social inequality, reinforce the rule of law and expand the social basis for economic growth over the years to come.

The other major challenge for democracy is the fight against organised crime and political violence. The consolidation of democratic regimes requires determined action against both these scourges, as well as corruption. Illegal drug trafficking plagues most of the region and it is uncertain whether there will be sufficient determination to curb this problem in the future. On the one hand, national anti-drug policies and massive US aid (in 2005, the US spent $724.5 million on anti-drug assistance to Latin America[10]) seem to have brought some results: a fall in the Andean cocaine output has been noticed, from 950 tons in 1996 to 645 tons in 2004.[11] On the other hand, some disturbing trends can be detected: the diversion of cocaine traffic routes towards Europe, the suspected spread of the plantations of coca in Peru and Bolivia, and the specialisation of Ecuador's criminal organisations in the transit of drugs and money laundering.[12]

Maintaining law and order in large conurbations where drug trafficking and other illegal trafficking activities are well established practices will be a huge challenge for local and national governments alike. On the other hand, one can be more optimistic regarding terrorism and civil strife. Terrorism in Latin America cannot be linked to the international Islamist networks and to the protracted Middle Eastern disputes that are fuelling these movements' resilience. Rather, terrorism in Latin America responds to home-grown perceptions and justifications. Terrorist attacks are essentially carried out by revolutionary movements.[13] The general trend in countries where terrorism has been present in the last decades is towards a weakening of the phenomenon, owing to a combination of police, military and political measures but also to the involvement of the US through its war against drugs and terrorism (mostly focused on Colombia). Most radical, pro-indigenous movements have opted for peaceful ways of protest, such as the Zapatista movement in Mexico.

Looking at these developments, it can be expected that, twenty years down the line, residual terrorism will no longer pose a threat to political stability. While the virulence of terrorist movements as such seems to be waning, containing widespread endemic violence will be of crucial importance for the development of the region. To give an order of magnitude of the damage inflicted on economies and societies, it has been estimated that in 1997 violence cost 10.5% of GDP in Brazil, 24.7% in Colombia, 24.9% in El Salvador, 12.3% in Mexico, 5.1% of GDP in Peru, and 11.8% in Venezuela.[14] The World Bank estimates that terrorist activities in Peru have caused a cumulated loss evaluated at $20 billion.[15] Since widespread urban poverty and unemployment are major contributing factors of violence, Latin America will face a serious governance and development challenge. Its economic growth demands a sound business environment to enhance private investment, but it also requires a good deal of social measures to reduce the inequalities and roll back the spread of violence.

External relations

Relations with the United States and region-building

Without any doubt the United States is the most important external actor in Latin America. However, whilst during the Cold War the United States maintained a keen political presence in the Western Hemisphere, in the last fifteen years most political developments have occurred without a direct US involvement. Since the years of the Clinton administration, the real issues in US-Latin American relations are of an economic, rather than political, nature. The creation of the North American Free Trade Area (NAFTA), the US reaction *vis-à-vis* the Mexican peso crisis in December 1994, and the launch of the Summit of the Americas the same year were illustrations of the US's change of attitude.

In Latin America, most observers welcome this hands-off policy, whereas some voices point out that US indifference (President Bush did not mention Latin America once in his State of the Union Address in January 2006) is not good news for the region. If the United States continues to focus its foreign policy on China and Asia, for commercial reasons, and in the Middle East region, due to security concerns and the energy reserves of that area, Latin America will have to get used to living with a northern neighbour who is not primarily interested in regional and local political problems. Another school of thought, though, suggests that a growing proportion of the United States population will come from Latin America, which will raise the American government's consciousness of internal problems in some countries.

Since the mid-1990s the 'international structures' in Latin America have consisted of three main trade blocs: NAFTA, Mercosur, and the Andean Community of Nations. These frameworks complement the all-encompassing regional body – the Organisation of American States. In addition, there exist two broad frameworks for transatlantic dialogue with Latin America: the EU-Latin America summits (the fourth meeting took place in Vienna in May 2006)[16] and the *Comunidad Iberoamericana de Naciones* (its fifteenth summit took place in Salamanca in October 2005), in which both Portugal and Spain are participating.

Against this background, new dynamics may be shaping the regional order. Peru's project to negotiate a free trade agreement with the United States – similar to the agreement between Chile and the US – has triggered criticism from Venezuela's President Chavez, who is proposing to reorganise the regional structures. Instead of a 'geographical logic', Chavez would like to introduce a 'political logic' whereby left-wing governments would create a 'peoples' association'. In spite of such proposals and despite current difficulties in the Doha round of the WTO, it can be expected that trade amongst Latin American countries, between these countries and the United States and between Latin American countries and the rest of the world, will increase in the years to come. Latin American interest in region-building efforts, with the aim of emulating the European integration process, will also increase.

Latin America, trade and globalisation

Latin American economies are relatively open to globalisation, and integrated into major trade flows. Intra-regional trade only represented 16% of total commerce in 2004.[17] If regional agreements and institutions are reinforced, intra-regional trade will likely grow. Today, most of Latin American trade is directed towards the European Union, which remains the first commercial partner of the region, and the United States. In 2004, the European Union was Brazil's first trade partner with more than 23% of its trade, ahead of the United States (20%) and Argentina (8.2%). Argentina's top trade partners were Brazil (23.4%), the European Union (21.1%), USA (12.7%) and Chile (7.3%). Finally, Chile's trade was conducted with the EU (19.8%), USA (13.9%), China (8.9%) and Argentina (8.1%). Mexico is a case apart: trade with the US accounted for 70.7% of the total, while the EU was Mexico's second trading partner with only 6.9%, ahead of China with 4.1%. Trade relationships with China are particularly dynamic across the region, sustained in particular by the strong Chinese demand for natural resources. China's imports from Latin America increased fivefold over the period 2000-2004.[18]

Twenty years on, it can be expected that Mexico will still be firmly linked to the US economy. Brazil will be a strong economic pole with robust ties to all its neighbours, particularly Argentina. The Andean countries will enjoy commercial relations with the two blocs. For its part, the European Union will maintain important commercial relations with major Latin American actors. It is not clear, however, whether Venezuela's wealth in natural resources will result in sustained economic development. Experience in resource-rich countries shows that dependence on the export of energy commodities is often a curse rather than a blessing.

From an economic and social standpoint, neither the Asian model nor the Washington consensus are magic formulas for fostering the development of Latin America. The Asian model cannot be applied for many reasons of a historical, cultural, social and political nature. The Washington consensus showed its limitations *vis-à-vis* Latin America during the 1990s, when extreme liberalisation and de-regulation measures triggered social and political crises in some countries. In future, those measures must be accompanied by an additional focus on social protection, the rule of law and democratic practices. Latin America will, therefore, need to find its own model of economic and political development, drawing from the recent experiences of countries as diverse as China, India, Ireland and Spain.[19] Latin America should learn from success stories across the world, where education, investment and openness have been crucial factors. It follows that this region can and should define its own path to development, between ultra-liberalism, on the one hand, and the populist and 'narcissist-leninist' temptations, on the other. Leaders such as former President Ricardo Lagos of Chile and Luiz Inacio Lula da Silva of Brazil seem to have found that happy medium. If Latin America is able to find its own path and keep on course, it will be able to exploit its considerable potential for growth.

Notes

1. See US National Intelligence Council, 'Latin America 2020: discussing long-term scenarios', Summary of conclusions of the workshop on Latin American trends, Santiago de Chile, 7-8 June 2004.
2. Were these growth projections to prove right, GDP per capita will grow but not particularly fast, compared to US levels in PPP. Argentina and Chile would be at 38% of the US level in 2020 (from respectively 33% and 30% in 2005), Brazil at 22% (from 21%) and Mexico at 23% (from 24%). See *Foresight 2020: Economic, Industry and Corporate Trends* (Economist Intelligence Unit, 2006).
3. It should be noted, however, that infrastructure coverage differs largely within the region. See Marianne Fay & Mary Morrison, *Infrastructure in Latin America & the Caribbean: Recent Developments and Key Challenges* (The World Bank, Washington D.C., 2005.)
4. Marianne Fay & Mary Morrison, op. cit.
5. *Global Development Finance: The Development Potential of Surging Capital Flows* (The World Bank, Washington D.C., 2006).
6. For these data, see Christophe-Alexandre Paillard, 'L'Amérique latine, nouvel acteur majeur du grand jeu énergétique mondial', *Défense nationale*, no. 4, April 2006.
7. Ibid.
8. However, the misuse of oil revenues by the Venezuelan government has diverted crucial financial resources from the oil sector and dissuaded international investors. As a result, the oil production has stagnated and has deprived the State of considerable budgetary inflows. Oil income has fallen from $49 billion in 2000 to $44 billion in 2003. See Gustavo Coronel, 'Oil, Bolivarist Fuel', *Heartland*, 1/2006.
9. These estimates are based on a definition of poverty as an income of less than $2 a day. Kathy Lindert, Emmanuel Skoufias and Joseph Shapiro, *Redistributing Income to the Poor and the Rich: Public Transfers in Latin America and the Caribbean* (The World Bank, Washington D.C., 30 March 2006). In addition, extreme poverty (people living with less than one dollar per day) hit 42 million people in 2002, or 9.5% of the population. Extreme poverty, however, is projected to decrease to 29 million in 2015 (6.9% of the population). See 'Global Economic Prospects: Economic Implications of Remittances and Migration' (The World Bank, Washington D.C., 2006).
10. National Drug Control Strategy, FY 2006 Budget Summary, February 2005.
11. Substitution of coca by other crops remains a challenge and some setbacks have recently been noticed (notably in Peru and Bolivia). According to the US National Drug Intelligence Center, 'wholesale-level drug distribution in the United States generates between $13.6 billion and $48.4 billion annually. Between $8.3 billion and $24.9 billion in drug proceeds is smuggled out of the United States by Mexican and Colombian DTOs [Drug Trafficking Organisations] across the US-Mexico border.' *National Drug Threat Assessment 2006*, National Drug Intelligence Center, January 2006, and *2005 World Drug Report*, Part 1, Office on Drugs and Crime (UNODC), United Nations.
12. See Global Workshop on Drug Information Systems, Activities, Methods and Future Opportunities, 3-5 December 2001, Vienna International Centre, Austria, United Nations, 2002; 2005 World Drug Report, op. cit.; and 'Battles won, a war still lost', *The Economist*, 10 February 2005.
13. The region is plagued by four major groups considered as terrorist by the US Department of State. Colombia counts three movements – the Revolutionary Armed Forces of Colombia (FARC), the National Liberation Army (ELN) and the United Self-Defense Forces of Colombia (AUC) – and the Sendero Luminoso is still active in Peru. Mark P. Sul-

livan, *Latin America: Terrorism Issues*, Congressional Research Service, RS21049, 18 January 2006.

14. Juan Luis Londoño and Rodrigo Guerrero, 'Violencia en America Latina: Epidemiología y Costos', (Banco Inter-Americano Del Desarrollo [Documento de Trabajo R-375], Washington D.C.,1999), cited by Paulo de Mesquitas Neto, *Crime, Violence and Democracy in Latin America*, Paper submitted for the Integration in the Americas Conference, 2 April 2002.

15. Robert Ayres, *Crime and Violence as Development Issues in Latin America and the Carribean* (The World Bank, Washington D.C., 1998).

16. See the European Commission Communication, 'A stronger partnership between the European Union and Latin America' 8 December 2005, COM (2005) 636 final.

17. If Mexico is excluded, intra-regional trade in Latin America accounts for 25% of its global trade. See *International Trade Statistics 2005* (World Trade Organization, 2005).

18. 'World Investment Report: Transnational Corporations and the Internationalization of R&D', United Nations Conference on Trade and Development, 2005.

19. Andrés Oppenheimer, *Cuentos chinos. El engaño de Washington, la mentira populista y la esperanza de América Latina* (Sudamericana, Buenos Aires, 2005).

The New Global Puzzle

The EU in Context

The EU in Context

The world in 2025: a snapshot

The review of global trends and of the evolution of pivotal regions and actors over the next twenty years delivers a contrasted picture of the world in 2025. A mix of continuity and discontinuity will characterise the development of the international system. Some fundamental trends, such as the globalisation of the economy, will continue and even intensify. This will enhance interdependence, but also magnify differences across and within states and regions. At the same time, the emergence of new global players will alter the balance of power and put global governance seriously to the test.

The world will be both more connected and more segmented than it is today. Twenty years from now, no overarching trend or actor seems likely to predominate and structure the international system. The growing complexity of the global environment will make it harder to exercise power and leadership, and to shape a global agenda to address shared challenges. The key to interpreting the future resides, therefore, not in grand designs or theories, but in assessing the interplay of countervailing trends and principles.

The structural factors shaping cooperation, competition and conflict over the decades to come will evolve. Leaving aside the traditional instruments of military and economic power, other factors will come to play a more important role than in the past. These will be of a material, intellectual and ideational nature including, respectively, natural resources (fossil fuels, water, arable land), knowledge (science and technological innovation) and legitimacy (in its political, legal and cultural dimensions). It follows that major global players will need to be able to deploy their external action across different spheres of influence with a view to both defending their interests and mastering global challenges together.

The global environment

Twenty years down the line, the global environment will be deeply affected by the interplay of demographic, environmental, energy and technological factors. On the whole, the direction of current trends points towards a deterioration in the living conditions of mankind. At the horizon 2025, unless unforeseeable (but not necessarily unlikely) disasters or political crises occur, the breaking point of demographic, environmental and energy sustainability will not yet have been reached. Our findings, however, indicate that by then the situation will be far more critical than it is today. Longer-term extrapolations to 2030 and 2040 show that many countries will be faced, in different ways, with severe environmental distress and growing shortages of water, food and energy. Effective policy decisions at the regional and global level, as well as massive investments in infrastructures and technological innovation (be it in renewable sources of energy or new biotechnologies to fight disease and malnutrition), will be crucial to avert clear and upcoming danger. Beyond a certain point of tolerance, the direct implication of worsening environmental conditions cannot but be political instability and spreading disorder.

In 2025, the population of the developing world is expected to grow from 5.2 to 6.66 billion, but to remain almost stable in the developed world (1.2 billion). The combined effect of massive demographic expansion and environmental deterioration in some regions is of great concern. This will notably be the case for most MENA countries and for vast areas of Sub-Saharan Africa. The population of MENA countries is expected to grow by around 40% by 2025, while the arid or desert areas, currently 87% of the whole region, will expand further because of higher temperatures and less rainfall. As a consequence, per capita water availability might shrink by half in 2025. While the picture is more varied in Sub-Saharan Africa, which features comparable rates of demographic growth, water stress or scarcity already affects a number of countries in the Sahelian strip, in the East and in the South. Given the impact of climate change, forecasts are negative. Arable land is being lost in other parts of the world as well, including China, where water scarcity, environmental degradation and pollution pose a serious threat to development over the medium term.

Pollution, urbanisation and industrialisation will represent a major environmental and health challenge for developing countries. Driven by the deterioration of the rural environment and/or by employment opportunities in cities, urbanisation rates will grow everywhere over the next twenty years, reaching 38% in India (550 million), 57% in China (over 800 million people), and 70% in the MENA region (380 million). In fact, people on the move within countries will largely outnumber international migrants. The high concentration of inhabitants in large urban centres or mega-cities across the developing world is already proving unsustainable. In these areas, human pollution is and will remain the biggest problem, while health services appear ill-prepared to contain spreading diseases and respiratory ailments. Worsening climatic conditions will compound these trends.

In the energy domain, trends point to a steep rise in greenhouse gas emissions, which will inevitably impact on climate change. Energy demand will grow by 50% between now and 2025, and two thirds of the additional demand will come from developing countries. Fossil fuels (oil, gas and coal), are expected to account for 81% of demand by 2025, with oil remaining the main source of energy for transport, and coal for power generation, notably in India and China. While considerable investments are envisaged by governments and private actors to enhance the application of sustainable renewables, their share of global energy consumption is expected to remain as low as 8% in twenty years' time.

Reflecting the projected evolution of energy demand and consumption, the emission of greenhouse gases is expected to increase at a fairly sustained pace. This is despite the envisaged efforts of developed countries to cut emissions, and the worldwide focus on enhancing energy efficiency. While forecasts are highly controversial, the current levels of concentration of CO_2 in the air might grow by 30% to almost 400% in 2100, depending on the scenario and on the extent of technological innovation. Over a shorter timeframe, by 2030, developing countries are likely to overtake developed countries in terms of absolute levels of emissions.

The rise in global temperatures seems therefore inescapable in the medium to long-term: it will essentially be a matter of slowing down this trend, and mitigating its effects. Preventive action will be crucial to avoid catastrophic effects over the longer-term. These would include not only global warming but also, probably, the cooling of the North Atlantic area due to the melting of the Arctic ice cap, which would upset oceanic streams and notably weaken the Gulf Stream. While the intensity and timeframe of this evolution remains a subject of scientific debate, what is clear is that ongoing deforestation, notably in South America and Asia, is generating additional emissions, while the absorption capacity of forest areas is shrinking correspondingly. Climate change will most likely intensify the frequency of floods, heatwaves and other natural disasters such as tropical hurricanes, whose destructive potential becomes clearer year by year.

In short, looking ahead to 2025, the world will be more populated, more exploited, more arid and more polluted than it is today. In other words, it may become a far less hospitable place. The wellbeing of billions of people will be put under more or less severe strain. Sound policy choices, however, can still prevent the aggravation of envisaged trends, and lead to a better management of natural resources. In particular, more efforts will need to be made to deliver 'global public goods' to those in need, whose absolute numbers will expand. These include, among others, access to health and education, and benefiting from a clean environment.[1] In other words, a viable long-term security strategy must focus on prevention across the different dimensions of external relations and reflect the notion of comprehensive security traditionally upheld by the European Union, and outlined by the European Security Strategy.

Globalisation and diversity

Globalisation is, and will remain, an important factor shaping international politics, economics and culture. That said, globalisation encompasses multiple dimensions, and entails contradictory implications. This will be all the more the case in the decades to come.

Economic globalisation will probably gain in speed and depth, with an ever-growing flow of goods, services and capital across a widening range of countries. The key difference from earlier stages of economic globalisation resides in the delocalisation and offshoring of manufacturing and services, empowered by new ICT technologies. This leads to integrated production processes at the global and regional level, whereby different countries provide different segments of the value chain.[2] From this standpoint, the world is expected to become more interdependent in the decades to come.

Economic globalisation has a mixed impact on global inequalities. From the (rather narrow) standpoint of per-capita income levels, it has resulted in a narrowing of the gap between emerging economies such as China, India and Brazil, and developed countries. This trend is expected to continue. On the other hand, those countries that fail to integrate in global trade and investment flows risk further marginalisation. Today, 25 countries account together for 80% of world trade, whereas 56 others represent less than 0.01% each.

In fact, our findings suggest that the implications of economic globalisation, and of the existing model of economic development, should be assessed against a broader range of parameters. The real living standards of people around the world will depend on more than relative wealth, although that remains an important indicator. Looking at the medium to long-term, the environmental and social costs of economic growth ought to be factored into the assessment of development and growth strategies. Otherwise, economic competition might soon undermine the sustainability of growth.

Pressures stemming from worldwide competition for markets and resources will put economies and societies under stress. Challenges, however, will of course differ for developed and developing countries (and within each of these subsets): there is no 'one-size-fits-all' response. Developed countries will need to adjust their economic systems so as to shift resources towards sectors of comparative advantage, with higher technological added value. Developing countries will need to address the inadequacies of political and economic governance as a matter of priority, and upgrade their infrastructures, so as to create an investment-friendly environment. Reform will entail tensions, and possibly draw new dividing lines within and between societies, between those benefiting from globalisation and those losing out from it.

While inequalities may be decreasing at the global level, the outlook for some critical regions, such as MENA and parts of Sub-Saharan Africa, is negative and seems to be getting worse. The envisaged deterioration of other structural factors, such as demography and the environment, could endanger the stability of those countries which are failing to adjust to globalisation and losing ground compared to emerging economies. Moreover, the perception of disparities, and the awareness of the challenges and threats looming ahead, will be magnified by the globalisation of information and perceptions.

Globalisation spills over from economics into politics and culture. Individuals and communities worldwide are affected by globalisation, which becomes a powerful determinant, among others, of political choices and social trends. This tendency will intensify in the future, as globalisation will be increasingly 'internalised in the day-to-day life of countries and societies.'[3] The cultural dimension of globalisation appears particularly contrasted.

The impact of growing media and ICT penetration will be twofold. While international media conglomerates and entertainment industries will continue to spread Western consumer values worldwide, new satellite channels, such as Al-Jazeera and Al-Arabiya, will open up the possibility of broadcasting different perspectives on reality.[4] At the same time, Internet and mobile communication are empowering the individual, opening up undisclosed potential for connectivity, influence, political initiative – and also organised violence. The global flow of ideas, information and images will boost interconnectedness. A more connected world, however, will not necessarily be more homogeneous. Growing interconnectedness is simultaneously exerting pressure towards homogenisation and fragmentation. The convergence of social habits around familiar patterns of consumer behaviour is paralleled by growing cultural fragmentation and, sometimes, alienation.

Globalisation will thus become less Westernised, more plural, more regional ('glocalisation')[5] and more hybrid:[6] cultural diversity is on the rise.[7] Future trends are likely to enhance both cultural contamination and confrontation. Societies will react differently depending on endogenous cultural, social, political and economic variables. Scientific and societal transformations, moreover, will continue to trigger value conflicts and debates on the ethical implications of scientific progress, notably in biotechnologies, and on how to live in an increasingly risky environment.[8]

From a political standpoint, the backlash against globalisation could take three main forms: political and economic nationalism, religious fundamentalism and grassroots protest. First, globalisation may fuel the resurgence of nationalism in inward-looking societies, including in some rich developed countries, with an emphasis on preserving or achieving political, cultural or ethnical homogeneity, and a tendency to reject 'the other'. Second, religious fundamentalist movements are expanding both in developing countries, notably in the Muslim world, and within Western societies. Where modernity poses a threat to peoples' traditional way of life, fundamentalism provides a firm ground with clear identities, clear principles and easy explanations.[9] Third, globalisation triggers various forms of protest and refusal, at both the domestic and international level. Non-state civil society actors, like NGOs and trans-national networks, play an important role in shaping the political agenda. Together with the increased awareness and mobilisation of public opinion, the cultural dimension of international politics will constrain the traditional power-holders in governments.

While being the dominant feature of the next two decades, globalisation will be neither entirely global (a number of countries will struggle to join the globalisation process, and possibly fail) nor total (within countries, part of the population will become, or will fear becoming, excluded). The pace of change is accelerating and the world is likely to become more interdependent and interconnected, but also more diverse and, in some regions, more unequal. Such a system is inherently unstable and hosts a considerable potential for tensions. In its interplay with local and regional dynamics, globalisation will lead to a more segmented world, with security dynamics varying greatly between different regions.

Three defining questions for the future

Political developments are inherently unpredictable, given the wide scope for human agency in shaping the course of events. At the same time, drawing from the development of structural trends and from the evolving strategic outlook of key international actors, some insights into the future international political system can be attempted. The purpose is not to try and anticipate specific events, but to sketch out the broad outlines of the global context to come, and point out the key questions concerning the future of power, legitimacy, and democracy. Four basic observations underline this perspective.

First, in the absence of a clear and established international political system, like the one based on the East-West confrontation during the Cold War, globalisation itself will be the most influential factor shaping international politics. The ways in which the new economic powers will be willing and able to translate their weight into some form of political influence will be key to the future political and security system.

Second, multipolarity will be a fact of life. The rise of new global and regional players such as China, India, Iran, Brazil and Indonesia among others will make the international system more heterogeneous. Third, global governance will be put under serious strain. The relationship between old and new powers will determine the future of global governance. A more interdependent and complex world generates challenges, which demand a coordinated response. At the same time, however, it might become more difficult to focus collectively on systemic issues such as poverty and environmental degradation.

Fourth, the ability of the West to influence international affairs will be put to the test as its share of world population and GDP is shrinking. The emerging powers bring with them their own vision of the world, which can differ considerably from that of the established ones. Consequently, the West will probably find it much harder to set the international agenda, and new ways of fostering international cooperation will need to be defined.

The future of the international system

Twenty years down the line, the international political system will be mixed, and will present three main features. First, the re-emergence of great power politics; second, the consolidation of regional multilateral frameworks, of which the EU will continue to represent the most advanced example; third, the potential proliferation of weak states. In some cases, weak states may fail to preserve the rule of law domestically, and fail to prevent disorder from spilling over their borders, thereby becoming sources of insecurity.

The co-existence of power politics, multilateral cooperation and unstable, marginalised countries or regions is not new in itself, but two novel elements can be observed. On the one hand, over the next twenty years, these three dimensions are likely to consolidate simultaneously, with no paradigm prevailing over the others (such as, for example, bipolarism largely did in structuring both international institutions and the pattern of peripheral conflicts during the Cold War). On the other hand, in an interdependent world, these dimensions will be more intertwined and overlapping than in the past. For example, strong powers may decide to make greater use of multilateral regional frameworks to enhance their influence, while violence spilling over from failed states might not only affect their neighbours but also fuel international terrorism and threaten major powers.

Within two decades, no major pole of power is likely to be hegemonic.[10] This is notably the case given the changing definition of 'power' which includes, well beyond the narrow military dimension, economic and cultural variables and, more generally, the ability to build international consensus. In other words, hegemony entails both material supremacy and the recognition of primacy and authority on the part of other members of the international community. The absence of a hegemonic power will have momentous implications for shaping global governance. The redistribution of power will lead to a redistribution of influence in setting the global agenda. Importantly, no individual state power will be up to the task of setting the rules of the game in the global economy, and shaping global institutions, as the US largely did in the aftermath of the Second World War.

A multipolar system is likely to emerge. The question is what type of multipolar system that will be. As new countries acquire power status, they may prove willing to mutually accommodate their interests so as to ensure the stability of the system and preserve their new prerogatives. That could be consistent with the consolidation of multilateral legal regimes, arguably the most effective platform to enhance cooperation.[11] Such a multipolar system would be of a relatively benign nature, along the lines of a traditional concert of (*status quo*) powers, underpinned by broad multilateral commitments.[12] The potential gridlock of international institutions, widening disparities and the emergence of a nationalist/protectionist discourse might, however, lead to a more conflictual form of multipolarism, with great powers competing for scarce resources, markets and spheres of influence.[13]

Developments in either of these directions will largely determine the future of the international system, with considerable implications for the international community's ability to confront common challenges, and for global security. The question will be whether and how the most powerful countries will be willing to address the structural problems of marginal regions. They will be faced with the challenge of how to foster and sustain over time the transition from endemic disorder to order of those countries or regions where the rule of law is absent. State failure and extreme poverty in peripheral regions may not pose a direct threat to the developed world, but is likely to undermine, over the long term, the broader requirements for the security of rich countries. In an interdependent world, disorder and insecurity will tend to spill over.

Concerning global security, the main question will be whether a collective and inclusive system of governance will prevail on the basis of multilateral norms, or whether a new kind of ideological bipolarity will emerge, opposing an alliance of democracies to the rest of the world. The first option would amount to reconciling a multipolar world and an effective multilateral system. The other alternative would be driven by the primacy given to Western interests and a more confrontational or defensive approach to the international system. The key test for the EU will be whether it will play a leading role in fostering the first option, thereby promoting its own distinctive values, and averting the danger of renewed ideological confrontation.

The future of global governance

Over the next twenty years, the demand for global governance will increase steeply. This will be due to two main factors: growing interdependence and growing heterogeneity. Globalisation demands governance: markets need law and regulations, stability and predictability. In an interdependent world, it will be increasingly hard to uphold the traditional boundaries between the different dimensions of global governance. Trade, finance, development aid, migration flows, energy supplies, environmental degradation, and new trans-national asymmetric threats will interact in shaping global dynamics.[14] The implications of economic globalisation, such as socio-economic imbalances and excessive stress on natural resources, require additional international focus as well. At the same time, the international system is growing more complex and heterogeneous. Global governance will need to adjust to a shifting balance of power and interests, and cope with the growing diversity of state and non-state actors.

Growing diversity between different state powers and also within domestic societies goes hand-in-hand with the multiplication of exchanges at all levels. The main challenge will be to define a sustainable balance between these two potentially conflicting trends. Due to technological progress, the economic and cultural dynamics of globalisation, and also international migration, heterogeneity and proximity are both deepening. This mix could entail serious tensions in the absence of sensible, collective management of resulting trends. Joint efforts will require both shared rules to establish what is acceptable and, therefore, legitimate for all key stakeholders in international affairs, and a progressive reform of global institutions.

Legitimacy will be the hard currency of future international relations, possibly the most important asset to ensure the long-term success of specific initiatives. It has been rightly pointed out that the notion of legitimacy is neither objective nor subjective, but 'inter-subjective'.[15] It follows that, as new pivotal players take the centre-stage, a wider range of interests and values will need to be taken into account to deliver viable compromise. While the search for agreement in defining the international agenda might well prove more complicated, it will also become more important. In confronting many of the challenges outlined in this Report, such as for example energy dependency, the choice will be between collective responsibility and collective irresponsibility. In other words, legitimacy will be a central component of power, or the ability to achieve appointed goals. This applies not only to the political, economic, environmental or cultural domains, but also to security and defence.

Needless to say, unilateral action will always be an option for individual countries, notably against direct and imminent threats. That said, success in countering trans-national terrorism or bringing order to areas of conflict will largely depend on the perceptions and judgment of the community of states and of global public opinion at large. It has been stressed that cultural factors, the narratives framing military action, and the values of intervening countries and of the countries where conflict unfolds, need to be internalised in strategic thinking for the legitimacy of military intervention to be acknowledged.[16]

By definition, international legitimacy requires multilateral consensus. International institutions at the global and regional level are the embodiment of multilateralism, and foster it through their output and their processes. There will be no effective multilateralism and, therefore, little international legitimacy on offer, in the absence of well-functioning international institutions. The institutional framework of global governance will need to undergo significant reform if it is to keep up with the pace of change, and manage risk. Failing that, the gap between the supply and the demand of global governance may well widen. It is increasingly open to question, however, whether the conditions enabling such reform are there.

The failure to accomplish a thorough reform of the United Nations in 2005 is a reason for serious concern, and a major failure of the international community. For all its shortcomings, the universal membership of the UN confers upon it unparalleled legitimacy. The progressive marginalisation of the UN, and notably of the UNSC, over the next 20 years cannot be excluded. This will essentially depend on whether major global players will pursue their interests unilaterally, or through more or less stable coalitions, as opposed to seeking consensus and strengthening the system of collective security at the global level.

The reform of global governance needs to pursue, in addition, better coordination between different international institutions. Steps in this direction can be envisaged in the field of development, towards the fulfilment of the Millennium Development Goals (MDG).[17] Comprehensive MDG-based poverty reduction strategies, focussed on capacity-building and long-term donors' commitments, are regarded as crucial to match the ambitious targets by 2015. Progress in this area needs to be closely monitored, given the link between poverty, disease and security concerns acknowledged in the European Security Strategy. A relation of direct proportionality exists between the level of GDP per capita in one country and its chances of lasting democratisation. On the other hand, the level of GDP per capita is inversely proportional to the probability of one country relapsing into conflict.[18]

The disconnect between different branches of international governance poses a problem for the future of international trade too, as the failure in closing the Doha Development Round testifies. Trade negotiations point to a widening North-South divide on the terms of management of global trade and the economy at large.[19]

Given the shortcomings of global institutions, different frameworks of governance may come to play a larger role. Trans-governmental networks may well emerge as a key feature of global governance, combining the required expertise to address complex technical issues with the direct involvement of national officials.[20] Yet another pillar of global governance may consolidate around the existing 'summitry' of national leaders. The G8 and the G20 have become important focal points for shaping the agenda of global governance.[21] Potentially, they combine the clout of national leaders with a degree of flexibility and the ability to establish issue linkages across sensitive, connected dossiers. The question is whether these summits will succeed in ensuring proper follow-up and implementation of their conclusions, and whether they should be regarded as a parallel framework to, or a potential competitor of, the UN system. Networks, whether at the political or technical level, lack the inclusiveness and the unique legitimacy of the UN, pose an accountability problem and entail the risk that competing networks may run conflicting agendas.

Regional organisations could be also complementary to global frameworks and/or partially compensate for their decline. Regionalism is a growing phenomenon, with bodies like the African Union, ECOWAS and ASEAN expanding their remit beyond the economic sphere to security matters. Inter-regionalism could gain further ground in the future, with the EU supporting regional partners (Asia-Europe Meeting [ASEM], and support to AU peace-keeping) and with larger regional forums such as the ASEAN regional forum (ARF) fostering dialogue across regions.[22] Looking at the challenges ahead, however, the potential contribution of regional organisations needs to be put in perspective. First, regional organisations are not effective in dealing with security crises. With the exception of the EU and NATO, their contribution is modest and requires support from the outside.[23] Second, it remains to be seen whether NATO and the EU will continue to provide the required support indefinitely, given resource constraints and vacillating domestic political will and commitment. Third, regional organisations may be the setting for hegemonic games, as in the case of the Shanghai Cooperation Organisation.[24] On the whole, it is hard to anticipate whether the potential of regionalism to support global governance will be realised, notably in the domain of security.

The future of democracy

Since the end of the Second World War and of the Cold War, the number of democracies in the world has been in constant expansion. Today, according to Freedom House, 122 countries are electoral democracies (64% of the total), as opposed to 66 twenty years ago (40% of the total). Less than 50% of existing states, however, are considered 'free' while 30% are defined as 'partly free' and 24% 'not free'.[25] In other words, while there is an obvious link between the two, democracy and freedom do not always go hand-in-hand. Successive waves of democratisation have raised new questions concerning the very definition of democracy, the complex dynamics of democratisation and the further expansion of democracy. At the same time, the future of democracy in the developed world is also subject to question.

The insightful distinction between 'illiberal' democracies and 'liberal' autocracies provides a useful guideline for assessing ongoing and future developments.[26] As a form of government, democracy is about the way in which government is appointed, by popular vote, and not the way power is exercised. The constitutional DNA of Western democracies is the distinctive set of principles and institutions of liberalism, with an emphasis on checks and balances and respect for individual rights. Political regimes can, however, be democratic while not being liberal. Conversely, some non-democratic regimes can guarantee the rule of law, proper services and a degree of respect for human and civil rights, while not permitting electoral competition challenging the ruling elite.

As illustrated in this Report, 'illiberal' democracies have proliferated over the last few years. These range from the 'managed or 'sovereign' democracy of Russia to the new populist democracies of South America. Hybrid regimes have emerged in Central Asia, and parts of the MENA region and Sub-Saharan Africa, where elections are most often than not an empty ritual to confirm those in power. On the other hand, it has been suggested that autocracies could well prove successful in preserving themselves, notwithstanding economic growth, by tightening their control on the media, the exercise of political rights and the education system.[27] On the whole, with the exception of European countries like Ukraine or Georgia, the agenda of democracy promotion seems to be plagued by old and new obstacles. New ways need to be devised to define an effective path of democracy promotion, free from ideological or missionary postures.

The question is whether political capital and incentives should be spent to promote electoral democracy in the first place, or whether more focus should be dedicated to the cultural, social and economic conditions enabling the progressive consolidation of the rule of law and, eventually, a solid liberal democratic regime. An overview of those regions of the world where democracy struggles to take root, from the Arab world to Sub-Saharan Africa and Central Asia, shows that socio-economic development and governance reforms are essential prerequisites for the emergence of sustainable democracy.

In fact, the superimposition of a 'democratic' form on a reluctant social, economic and cultural 'structure' can rarely be sustained. Pre-existing social structures and networks need to be co-opted into the process of democratisation, otherwise there is a strong risk of creating powerful opponents to the process itself. The economy needs to develop in such a way that workers and entrepreneurs have a real stake in market freedoms and the rule of law, and call for more political accountability in exchange for taxation. On the whole, therefore, future efforts at fostering democracy should be grounded in a clear understanding of local history, culture, and governance regimes.

In addressing the question of democracy promotion, the trade-off between democracy and stability will come into play. First, weak democracies, unable to impose the authority of the state on their territory, and/or perceived as illegitimate by parts of the population, could degenerate into failed states and spread instability. Second, democracies are not necessarily peaceful. Unless they are reinforced by strong constitutional constraints, new democracies are fertile ground for populist leaders, prone to consolidate their power by mobilising the masses with a nationalist, aggressive rhetoric.[28] This was, for example, the case of in the former Yugoslav countries in the early 1990s.

In perspective, serious challenges confront democratic regimes in the West as well. The control exerted by private interests on policy-making and the corresponding loss of prestige and influence of traditional political parties, particularly noticeable in the US but significant across Europe too, undermine an open public debate and the definition of collective preferences. The growing personalisation of politics goes hand-in-hand with the progressive strengthening of the executive, notably in relation to the legislative branch. Opinion-poll driven politics pose a serious problem of leadership, with political classes increasingly inclined to bend to the demands of the majority as opposed to taking alternative, albeit unpopular, courses of action. In this connection, the growing share of the elderly population in Europe might lead to an increasingly conservative and defensive political discourse.

Many of these trends are symptoms of a larger, deep-rooted malaise, which can be interpreted as the 'securitisation' of the political discourse or the 'politics of survival'.[29] A widespread sense of insecurity and uncertainty pervades Western public opinion (although to different degrees in different countries) and affects the perception of existing and future threats and challenges, from environmental degradation to energy shortages. Transnational terrorism poses a particularly serious challenge because of its ability to strike at any time anywhere, potentially employing weapons of mass destruction. The feeling that danger exists within Western societies, as opposed to faraway places, may change the collective perception of the threat.

This evolution is particularly insidious when linked up to the challenges of multiculturalism, notably in Europe. Struggling integration policies, social and economic inequalities and the sometimes violent contraposition of alternative values systems represent a dangerous mix. The ongoing 'securitisation' of the debate on migration and integration might, in perspective, generate the mutual alienation of different groups in our societies. Extreme political formations – with xenophobic populist movements on the right of the spectrum and anti-global and anti-market groups on the left – are favoured by these developments, fuelling populist and/or protectionist feelings.

The core of the matter is that public confidence in the ability of national political systems and establishments to confront upcoming challenges and ensure progress is declining. Were these trends to deepen, the liberal foundations of democratic regimes in the West could be challenged. This would be all the more liable to happen were some of the most dreaded threats to materialise, including major terrorist attacks or the eruption of pandemics, which would conceivably strengthen the hand of the state in crisis management. Ethical and political dilemmas are likely to emerge in the future, which could be usefully anticipated in the current political debate.

The European Union in context

Internal challenges

In 2025, the EU is likely to be still one of the richest and safest parts of the world. However, in several areas, current trends indicate that the EU might have difficulties in maintaining its status. In many Member States, population ageing will dramatically change old-age-dependency rates. This will imply high costs for medical and social care and will have a significant impact on public finances. Many European industries are likely to remain competitive, but they will continue to offshore important parts of their production (and development). Value will thus increasingly be generated outside the EU.

Europe will probably remain at the cutting edge in key technology areas such as telecommunications or renewable energies. In other areas, its position may be less certain, either because competitors already have a head start (e.g. the US in IT and nanotechnology) or have boosted their research investments at an increasingly faster rate and are catching up rapidly (e.g. India in IT, China in biotechnology and nanotechnology).

Far-reaching reforms of social security systems, labour markets and education systems will be necessary to modify these trends or at least to cope with their consequences. Faced with the challenges of globalisation, many Europeans feel less secure, in economic, cultural and physical terms than in the past and are losing confidence in public institutions. The challenge will be to foster economic competitiveness without endangering social cohesion. The outcome will be decisive both for Europe's internal stability and its capacity to generate the necessary resources to play a serious role in world affairs.

In many areas (including for example research and development, energy, and immigration), closer cooperation and/or integration at the EU level would help to improve policy output. However, whether greater competition on the global level will drive Member States closer together or apart remains to be seen. European integration is often perceived by national public opinions as an additional threat to national interests and identities rather than a means of coping with the challenges of globalisation. Widespread sentiments of fear and defensiveness *vis-à-vis* globalisation represent a perfect breeding ground for political populism and economic protectionism, which can also hamper further European integration (and/or enlargement). Social polarisation and population ageing can reinforce such tendencies in the future even more.

Regarding its energy supply, the EU will find itself in an uncomfortable strategic situation. In 2025, OPEC, and in particular Saudi Arabia, Iran, Iraq and Algeria, are expected to provide around half of the EU's oil needs, while the remainder will come from Norway and Russia. For gas, Russia will certainly remain the main supplier, followed by Algeria and Norway. Given the geographical distribution of the world's energy resources, and the expected growth in demand, scope for diversifying the supply base is limited. This means that the Union's energy supply will almost completely depend on imports from regions with a high potential for instability. Even a temporary energy cut-off would have a major economic impact on the EU and directly affect its strategic interests.

A turbulent neighbourhood

The EU is surrounded by regions with a high potential for problems and tensions. In many countries in these regions, long-term trends are far from promising. In 2025, the full impact of most of these trends will not have been felt yet but, by then, the political, economic, and security outlook at the borders of the Union is likely to have deteriorated considerably, even if catastrophic shocks (new epidemics, natural disasters, major wars) do not occur.

Russia's economy is growing at a fast pace and the central government has reasserted its authority, but the transition of Russia towards a stable democracy and market economy remains very much under question. A great number of uncertainties plague its future development and may lead the country to follow quite a different trajectory over the next two decades. Economic growth will continue to suffer from the persistent lack of diversification, an underdeveloped private sector, and dependency on energy exports. Russia's population will continue to shrink, with low fertility rates and life expectations. Social problems are on the rise (poverty, alcoholism, crime) and there are growing disparities between different social groups and between different regions, with wealth concentrated in some cities. These unbalances and the poor standards of living across the country open the door to renewed nationalism and xenophobia. In parts of Russia and of the post-Soviet space, moreover, Islamic fundamentalism could pose a bigger, destabilising threat in the future.

At the international level, Russia will seek to re-acquire its great power status by playing on different tables, with different partners. Energy, in particular, is likely to remain a strategic asset for Russia's foreign policy, and attempts to diversify its exports network can be expected, with potentially disturbing implications for Europe. The military will receive more attention and, economic growth permitting, funding, and will remain a key instrument for Russia to assert its interests along its turbulent Southern flank. Compared to its relationships with the US and China, Russia's focus on developing a strategic partnership with the EU could gradually decline.

The Middle East will remain a conflict-prone region, riven by deep-running political and ethnical tensions. The endemic conflicts in Palestine and Iraq are likely to remain factors of systemic instability for the foreseeable future. The Iranian question will be a key variable for the regional security equation. In addition, problems are accumulating in a number of areas. Water scarcity and rapid urbanisation will lead to a deterioration in environmental conditions and living standards. Many countries are experiencing great difficulties in reforming their economies and will probably be unable to provide enough jobs for millions of young people entering the labour market over the next 20 years. Unemployment and social inequalities are likely to increase and become the breeding ground for anger, alienation and unrest, further strengthening fundamentalist Islamic movements as alternatives to authoritarian regimes. Domestic governance will remain a major concern. In some countries (such as Iraq and potentially Lebanon), ethnic conflicts and the growth of sectarianism may continue to fuel political tensions and undermine democratisation. In others (Saudi Arabia, Egypt), political transition towards less repressive regimes may cause major unrest and lead to considerable instability in the region.

The possibility of a systemic breakdown of the Middle East, along the so-called 'crescent of crisis', cannot be ruled out.[30] That could imply state failure, the downfall of pro-western, authoritarian regimes and an even closer, perverse connection between different conflicts across the region (Israel-Palestine, Iraq, Afghanistan, Chechnya, Kashmir, Lebanon, Kurdistan). Even in the absence of a major crisis, regimes in the Middle East are likely to remain less than reliable partners for Europe. This will notably be the case if Europe's ambivalence on how best to promote good governance and engage both with reformists and (non-violent) Islamist conservatives continues.

Sub-Saharan Africa will probably remain affected by both traditional and new scourges, although some signs of progress and reform can be detected across the continent, and notably in the Southern part. In many regions, environmental degradation is likely to worsen, due in particular to water scarcity, over-farming and unbalanced urbanisation. High fertility rates will lead to continuous population growth, but persistent poverty and diseases are likely to keep life expectancy low. Most economies will remain focused on agriculture, with primary commodities as the only relevant exports on world markets. Some countries may perform better than others, but the continent as a whole is expected to continue to decline relative to other developing regions. Brain drain is likely to further weaken Africa's human capital and reduce chances for economic and political reforms. In a number of countries, bad governance and ethnic conflicts may further undermine the faltering authority of the state, leading to its potential dissolution. In other cases, it is the way in which strong states exert their power that triggers domestic instability and serious conflicts.

The potential plunging of swathes of Africa into even deeper poverty, famine and conflicts may not pose a direct threat to Europe but will no doubt affect the broader requirements for its security. State failure may create lawless areas where militias, criminal organisations and terrorists thrive. Migration patterns from the region, projected to remain stable, might undergo sudden alterations with unpredictable peaks. The provision of oil and other raw materials might also be disrupted. At a different level, the biggest challenge to the credibility of Europe's claim to effectively support regional integration and good governance might come from tangible failure in this region.

The challenge for the EU: shaping change

Global developments will entail fundamental changes to the distribution of resources and influence. In other words, after three centuries of Western hegemony, history is taking a somehow more natural course, with new pivotal players emerging and smaller countries forging closer ties at the regional level. Power will inevitably shift from the West to the rest.

This process does not necessarily pose a threat, but it needs to be reckoned with for the very purpose of managing change in a balanced and peaceful way. While relatively smaller in demographic and economic terms, and exceedingly dependent on energy supply from its near abroad, the EU will not necessarily be weaker twenty years down the line. It will remain a major economic power, and its standards of living will continue to go unmatched, except by other rich countries such as the US and Japan. There is nothing fatalistically preordained in the shape of the world to come, and in the EU's place therein. It is a matter of political decision, drawing on Europe's comparative strengths.

▶ The first challenge facing the EU is to drive, as opposed to enduring, change. The biggest strength of Europe is its own experience of continental integration and stabilisation, and the new language of international relations that it has generated and spread worldwide. The biggest weakness is that Europe falls short of embedding its unique experience and normative orientation into a solid bedrock of shared interests, consistently defined and pursued.

The identification and formulation of shared interests, as the basis for common positions on key global issues, is a highly political exercise. The EU can hardly be expected to bring further added value if its Member States refrain from engaging in serious political debate on what they want to do together, and if they are reluctant to put their money and resources where their mouth is. That requires first and foremost that national political establishments change gear, and switch their discourse on foreign and security policy from a national to a European level.

Looking at the major economic, energy, environmental and demographic trends shaping the world to come, it is apparent that not making collective choices at the European level effectively amounts, over the long-term, to abdicating sovereignty. From this standpoint, the notion of 'European sovereignty' as complementary to national sovereignty should be taken as a reference for the debate. In tomorrow's international relations, the traditional notion of national sovereignty - as the absence of superimposed authority to that of the state on the people living in its territory - seems sterile. Collective action at the EU level would, on the contrary, be compatible with a positive, contemporary definition of sovereignty as the ability to effectively achieve results in cooperation with others.

▶ The second challenge for the EU is to foster the convergence between a multi-polar political system and a multilateral order, making them not only compatible but also complementary. Following the review of the future strategic outlook of major global players such as the US, Russia, China and India, and of the major trends presented in this Report, two considerations stand out. On the one hand, there is little doubt that the heavyweights in the international system do not and will not regard international norms and institutions as an end in themselves, but as one of the means to pursue their interests. As such, multilateralism *à la carte* might well remain, for the foreseeable future, the favourite option for established and emerging powers. On the other hand, when focussing on long-term global trends and their implications, multilateral solutions will prove the most effective to uphold the common interests of all stakeholders.

The larger a state power is, the more it is likely to benefit from effective multilateral norms and procedures in all domains. Likewise, the multilateral order will be all the more effective if backed by strong, respected stakeholders. If it ever existed as a sustainable option in the past, therefore, the contraposition between the calculation of national interests and the reliance on international institutions is likely to fade. Multilateral norms will probably acquire more prominence as a key variable in the definition of the strategic interests of large state actors. In a more interconnected world, the more 'connected' states will be those standing the best chance of profiting from globalisation, extending their influence and preserving their security.

Given the growing diversity and heterogeneity of the international system, one of the hard currencies of tomorrow's international relations will be legitimacy, in the eyes of both governments and average citizens across the world. Interdependence will generate increasing demands for common policies at the regional or global level, for example to manage energy resources, the environment and migratory flows. But the same applies to the resolution of conflicts and the preservation of peace. The supply of these policies, and the delivery of global public goods, will crucially depend on the ability to find compromise. The latter will be facilitated by multilateral institutions at the global and regional level.

Promoting effective multilateralism does not mean, therefore, neglecting power and interests, but channelling them through stable institutional channels towards sustainable, shared solutions at the international level. The solution to global problems lies in a new grammar of international relations, where the traditional concepts of power, interest and sovereignty are not abolished but revisited. The EU is a frontrunner of this process and has therefore a comparative advantage in defining and exporting the principles of institutionalised, norm-based multilateralism.

▶ With a view to rising to this opportunity, the third challenge for the EU will be to adapt to the new global context. This Report shows that the world in 2025 is likely to be more interdependent, complex and unstable. This makes it all the more necessary that EU Member States opt for cooperation and integration and manage to speak with one voice, and deliver the same message, in international affairs. The ongoing debate on reforming the European policies and institutions must be reviewed in light of the future challenges the EU will have to face, and not only of past and present controversies. No doubt, managing a larger Union, due to expand further, proves complicated and absorbs a lot of intellectual energy and political capital. The real challenges for Europe's future prosperity and stability, however, lie beyond its borders, from geopolitical tensions in neighbouring areas to the impact of emerging global players on international politics and economics. In other words, the debate on EU reform should go hand-in-hand with a strategic reflection on the values, interests and goals of the EU in international affairs. By thinking strategically of its own future and of its position in the world, the EU will be better equipped to reform itself in order to shape a more secure and better world.

Notes

1. Sven Biscop, *The European Security Strategy: A Global Agenda for Positive Power* (Ashgate, Aldershot, 2005).
2. Thomas Friedman, *The World is Flat: A Brief History of the Twenty-first Century* (Farrar, Straus and Giroux, New York, 2005).
3. The future of the WTO – Addressing institutional challenges in the new millennium', Report by the Consultative Board to the Director-General Supachai Panitchpakdi, World Trade Organisation, 2004.
4. The different press coverage of the Iraq war (CNN versus Al Jazeera) is certainly one example. For an overview of global media trends see: *Globalization of the Media Industry and Possible Threats to Cultural Diversity*, Final Study, European Parliament – Directorate General for Research, July 2001.
5. An amalgam of 'globalisation' and 'localisation'. The term was originally used to emphasise that the globalisation of a product is more likely to succeed when it is adapted to each local culture in which it is marketed. See: Mike Featherstone, Scott Lash and Roland Robertson, *Global Modernities* (Sage, London, 1995).
6. Philippe Legrain, 'Cultural Globalization is not Americanization', *The Chronicle*, 9 May 2003. Radley Balko, *Globalization and Culture: Americanization or Cultural Diversity?*, April 2003. (www.globalpolicy.org/globaliz/cultural/2003/03american.html). See also: Wendy Griswold, *Cultures and Societies in a Changing World* (Sage, London, 2003). According to Griswold, 'cultural purity is gone from the face of earth, it was probably always a myth, but now few even pretend to believe in it. We are all hybrids now.'
7. UN Human Development Report 2004, *Cultural Liberty in Today's Diverse World*. See also: Mike Featherstone, *Global Culture: Nationalism, Globalisation and Modernity* (Sage, London, 1990).
8. Ulrich Beck, *Risk Society: Towards a New Modernity* (Sage, London, 1992).
9. See: Martin Marty and Scott Applebee, *Fundamentalisms and Society: Reclaiming the Sciences, the Family, and Education* (University Of Chicago Press, Chicago, 1997), and Gregory F. Treverton et al., *Exploring Religious Conflict* (RAND, 2005). According to Marty, the major characteristics of Fundamentalists are as follows: (a) their religion forms their identity; (b) there is only one truth; (c) they are purposefully shocking; (d) they see themselves as part of a cosmic struggle; (e) they interpret historical events as part of this struggle; (f) they declare any opposition to themselves bad and immoral; (g) they only emphasize some parts of their heritage; (h) their leaders are typically male; (i) they rebel against the political establishment.
10. Niall Ferguson, 'A World Without Power', *Foreign Policy*, July/August 2004. See also, G.J. Ikenberry, 'A Weaker World', *Prospect*, November 2005.
11. John Gerard Ruggie, *Multilateralism Matters: The Theory and Praxis of an Institutional Form* (Columbia University Press, New York, 1993); Stephen D. Krasner (ed.), *International Regimes* (Cornell University Press, New York, 1983).
12. Hubert Védrine, *Face à l'hyperpuissance: textes et discours 1995-2003* (Fayard, Paris, 2003) and Charles Kupchan, *The End of the American Era: US Foreign Policy and the Geopolitics of the Twenty-First Century* (Random House, New York, 2002).
13. John J. Mearsheimer, *The Tragedy of Great Power Politics* (W.W. Norton and Company, New York, 2001).
14. Michael Dillon, *Global Security in the 21st Century. Circulation, Complexity and Contingency*, Chatham House Briefing Paper 05/02, October 2005.
15. Ward Thomas, 'La légitimité dans les relations internationales: dix propositions', in

Gilles Andréani, Pierre Hassner (eds.), *Justifier la guerre? De l'humanitaire au contre-terrorisme* (Presses de Sciences Po, Paris, 2005).

16. Lawrence Freedman, 'The Transformation of Strategic Affairs', IISS *Adelphi Paper* 379, 2006.
17. *Investing in Development, A Practical Plan to Achieve the Millennium Development Goals*, UN Millennium Project, directed by Jeffrey D. Sachs, 2005. For information on Millennium Development Goals, see http://www.un.org/millenniumgoals/.
18. Following the cessation of hostilities, the probability of a new conflict within five years is as high as 15% in countries with a GDP per capita of US$250 and goes down to 1% in countries whose per capita GDP is around US$5,000. See the *Human Security Report* 2005 (Human Security Centre, The University of British Columbia, 2005).
19. David Held, *Global Covenant, The Social Democratic Alternative to the Washington Consensus*, (Polity Press, Cambridge, 2004). See also the related debate on 'collective preferences' that extend beyond purely trade matters to affect basic social and political choices at the basis of global governance. Pascal Lamy, *The Emergence of Collective Preferences in International Trade: Implications for Regulating Globalisation* (PROSES, Sciences Po, 2004). Enhanced horizontal coordination between the World Trade Organisation and Bretton Wood institutions falls short of what is required for a coordinated response to trade-related assistance and capacity building. See the Integrated Framework of Cooperation between the secretariats of the IMF, ITC, UNCTAD, UNDP, WB and WTO, facilitating trade-related assistance and capacity-building needs of less developed countries. Referred to in *The Future of the WTO, Addressing Institutional Challenges in the New Millennium*, Report by the Consultative Board to the Director General Supachai Panitchpakdi, 2004.
20. Anne Marie Slaughter, *A New World Order* (Princeton University Press, 2004). Networks are active in domains as diverse as global financial architecture, antitrust rules, intellectual property rights, and international labour standards. See also Jean François Rischard, *High Noon: 20 global issues and 20 years to solve them* (Perseus Book Group, New York, 2002).
21. John Kirton, *The Future of the G8*, paper submitted at the Gleneagles summit, 2005; John Kirton, *From Collective Security to Concert: The UN, G8 and Global Security Governance*, 2005; John Kirton, *From G7 to G20: Capacity, Leadership and Normative Diffusion in Global Financial Governance*, 2005.
22. See two important collections for hints in this direction: Mario Telo (ed.), *European Union and New Regionalism* (Ashgate, Aldershot, 2001) and Mary Farrell, Björn Hettne and Luk Van Langenhove (eds.), *Global Politics of Regionalism* (Pluto Press, Cambridge, 2004).
23. Richard Gowan, *The EU, regional organisations and security: strategic partners or convenient alibis?*, Egmont Paper 3, IRRI, 2004.
24. Benjamin Goldsmith, 'Here there be Dragons: The Shanghai Cooperation Organisation', Centre for Defence Information, 2005.
25. *Freedom in the World 2006* (Freedom House, 2006).
26. Fareed Zakaria, *The Future of Freedom: Illiberal Democracy at Home and Abroad* (W.W. Norton & Company, New York, 2003).
27. Bruce Bueno de Mesquita and George Downs, 'Development and Democracy', Foreign Affairs September/October 2005.
28. Fareed Zakaria, op. cit.
29. Marc Abélès, *Politique de la survie* (Flammarion, Paris, 2006).
30. Ivo Daalder, Nicole Gnesotto and Philip H. Gordon (eds.), *Crescent of Crisis: US-European Strategies for the Greater Middle East* (Brookings Institution Press & EU Institute for Security Studies, 2006).

The New Global Puzzle

Annexes

Bibliography

The bibliography includes a selection of the sources that have been consulted to compile this Report. The sources have been divided into different lists according to the section for which they are most relevant, in order to facilitate access to them. The bibliography of Part I – The Trends – is broken down into the five main driving factors outlined in the Report. These lists notably include references to major publications from the UN, the EU, The World Bank, the International Energy Agency and other international organisations. The bibliographies of Part II – The Regions – and of the final section of the Report – The EU in Context – are grouped together separately. Many of the articles, working papers and books that featured as important sources in our work are listed in this second part of the bibliography.

PART I: The Trends

1. DEMOGRAPHY

- Alexandratos, Nikos. *Countries with Fast Growing Population in Midst of Global Demographic Slowdown: Issues of Food, Agriculture and Development*, Working Paper, Food and Agriculture Organization, Rome, March 2005.

- Baldwin-Edwards, Martin. *Migration in the Middle East and Mediterranean*, Regional Study prepared for the Global Commission on International Migration, Mediterranean Migration Observatory, University Research Institute for Urban Environment and Human Resources, Panteion University, January 2005.

- Bongaarts, John & Bulatao, Rodolfo A. (eds.). *Beyond Six Billion: Forecasting the World's Population* (National Academy Press, Washington, D.C., 2000).

- Caldwell, John C. *The Contemporary Population Challenge*, Paper Presented to Expert Group Meeting on Completing the Fertility Transition, Population Division, United Nations, New York, 11-14 March 2002.

- Center for Strategic and International Studies. *'Naming Names': Ageing and Vulnerability*, CSIS Program, Issue 3, 2 December 2002.

- Center of Health and the Global Environment. *Climate Change Futures: Health, Ecological and Economic Dimensions*, Harvard Medical School, November 2005.

- Central Intelligence Agency (CIA). *Growing Global Migration and its Implications for the United States (2000-2015)*, National Intelligence Council (NIC) Papers, United States, 2001.

- Coleman, David A. *Who's afraid of low support ratios? A UK response to the UN Population Division Report on 'Replacement Migration'*, Paper prepared for the United Nations Expert Group meeting, New York, 2000.

- Council on Foreign Relations and Milbank Memorial Fund. *Addressing the HIV/AIDS Pandemic: A US Global AIDS Strategy for the Long Term*, May 2004.

- Degli Innocenti, Nicol. 'Virus Hits at the Country's Life Force', *Financial Times*, 26 September 2001.

- Derr, Mark. 'New Theories on the Black Death', *International Herald Tribune*, 4 October 2001.

- Economic and Social Council. *World Population Monitoring, Focusing on International Migration and Development*, Report of the Secretary General, United Nations Commission on Population and Development, 25 January 2006.

- Ehrlich, Paul R. *The Population Bomb* (Buccaneer Books, New York, 1997).

- European Commission. *Confronting Demographic Change: A New Solidarity Between the Generations*, Green Paper, COM (2005) 94 Final, 16 March 2005.

- European Commission. *The impact of ageing on public expenditure: projections for the EU25 Member States on pensions, health care, long-term care, education and unemployment transfers (2004-2050)*, Special Report no. 1, The Economic Policy Committee and The European Commission, 14 February 2006.

- Falcam, Leo A. 'An Early Warning by Pacific Islands to the Mighty', *International Herald Tribune*, 16 August 2001.

- Faruqee, Hamid & Mühleisen, Martin. 'Japan: Population Ageing and the Fiscal Challenge', *Finance and Development*, IMF, vol. 38, no. 1, March 2001.

- Feld, S. 'Active Population Growth and Immigration Hypotheses in Western Europe', *European Journal of Population*, no. 16, 2000.

- Feshbach, Murray. 'Dead Souls', *Atlantic Monthly*, January 1999.

- Fourth World Water Forum. *African Regional Document: Water Resources Development in Africa*, Mexico, 16-22 March 2006.

- Friedman, Thomas. *The Lexus and the Olive Tree: Understanding Globalization* (Farrar, Straus & Giroux, New York, 1999).

- Gallup, J.L. & Sachs, J.D. *The Economic Burden of Malaria*, CDI Working Paper 52, Centre for International Development, Harvard University, Boston, 2000.

- Glenn, Jerome C. 'The World in 2025', *The Globalist*, 14 October 2001.

- Grant, Jonathan, et al. *Low Fertility and Population Ageing: Causes, Consequences, and Policy Options* (MG-206-EC, Rand Corporation, Santa Monica, 2004).

- Grant, Lindsey. 'Replacement Migration: the United Nations Population Division on European Population Decline', *Population and Environment*, vol. 22, no. 4, 2001.

- Harrison, Paul. *AAAS Atlas of Population and Environment*, American Association for the Advancement of Science (University of California Press, Los Angeles, 2001).

- Heymann, David. *The Urgency of a Massive Effort Against Infectious Diseases*, World Health Organization, 2000.

- Hille, Hubertus & Straubhaar, Thomas. 'The Impact of the EU Enlargement on Migration Movements and Economic Integration: Results of Recent Studies', in *Migration Policies and EU Enlargement: The Case of Central and Eastern Europe*, Organisation for Economic Co-operation and Development, Paris, 2001.

- Huntington, Samuel. 'Migration Flows Are the Central Issue of Our Time', *International Herald Tribune*, 2 February 2001.

- Inada, Masakazu, Kozu, Takashi & Sato, Yoshiko. 'Demographic Changes in Japan and their Macroeconomic Effects', Report no.04-E-6, Bank of Japan, September 2003.

- Just, Tobias & Korb, Magdalena. *International Migration: Who, Where and Why?* in Current Issues: Demography Special, Deutsche Bank Research, Frankfurt, 1 August 2003.

- La Documentation française. « La Grande Menace Démographique », *Problèmes économiques, Numéro spécial*, no.2.656-2.657, 2000.

- Livi-Bacci, Massimo. *A Concise History of World Population* (Blackwell Publishers, Oxford, 1997).

- Lutz, Wolfgang, et al. 'China's Uncertain Demographic Present and Future', Interim Report IR-05-043, International Institute for Applied Systems Analysis, Austria, 5 September 2005.

- Lutz, Wolfgang, Sanderson, Warren & Scherbov, Sergei. 'The End of World Population Growth', *Nature*, no. 412, 2 August 2001.

- Lutz, Wolfgang, Vaupel, James W. & Ahlburg, Dennis A. (eds.). 'Frontiers of Population Forecasting', *Population and Development Review*, A Supplement to Volume 24, 1998.

- Macura, Miroslav, MacDonald, Alphonse L. & Haug, Werner. *The New Demographic Regime: Population Challenges and Policy Responses*, UN Economic Commission for Europe and UN Population Fund, 2005.

- McNeil, Donald. 'Demand for Antibiotic May Alter US Patent Policies', *International Herald Tribune*, 18 October 2001.

- Meyerson, Frederick A.B. 'Replacement migration: a questionable tactic for delaying the inevitable effects of fertility transition', *Population and Environment*, vol. 22, no. 4, 2001.

- Misikhina, S., Pokrovsky, V., Mashkilleyson, N. & Pomazkin, D. 'A model of social policy costs of HIV/AIDS in the Russian Federation', Research and Policy Analysis, International Labour Office, Geneva (undated).

- Monitoring the AIDS Pandemic (MAP) Network. *AIDS in Asia: Face the Facts, A Comprehensive Analysis of the AIDS Epidemics in Asia*, Washington D.C., 2004.

- Monnier, Alain. 'Les évolutions démographiques de l'Union européenne', *Questions internationales*, no.18, La Documentation française, Paris, March/April 2006.

- National Intelligence Council. *The Next Wave of HIV/AIDS: Nigeria, Ethiopia, India, Russia and China*, Washington, D.C., September 2002.

- National Intelligence Estimates. *The Global Infectious Disease Threat and Its Implications for the United States*, NIE 99-17D, January 2000.

- Organisation for Economic Co-operation and Development. *Reforms of an Ageing Society*, Paris, February 2001.

- Organisation for Economic Co-operation and Development. *Trends in International Migration*, SOPEMI (Système d'observation permanente des migrations) Report, 2004.

- Passel, Jeffrey S. *Estimates of the Size and Characteristics of the Undocumented Population*, Pew Hispanic Center, Washington D.C., 2005.

- Peterson, Erik R. 'SARS, Lessons for the Longer Term', Center for Strategic and International Studies, Issue 9, April 2003.

- Peterson, Peter G. *Gray Dawn: How the Coming Age Wave Will Transform America and the World* (Crown Publishing Group, New York, 2000).

- Postrel, Virginia. *The Future and Its Enemies* (Free Press, New York, 1998).

- Regling, Klaus. 'How Ageing Will Torpedo Europe's Growth Potential', *Europe's World*, Spring 2006.

- Rincon, Paul. 'Faster emergence for diseases', *BBC News*, 20 February 2006.

- Roudi-Fahimi, Farzaneh, Creel, Liz & De Souza, Roger-Mark. *Finding the Balance: Population and Water Scarcity in the Middle East and North Africa*, Population Reference Bureau, Washington D.C., July 2002.

- *The Economist*. 'Life Story', 29 June 2000.

- The Global Fund to Fight Aids, Tuberculosis and Malaria. *HIV/AIDS, Tuberculosis and Malaria: The Status and Impact of the Three Diseases*, Geneva, 2005.

- *The Globalist*. 'The European Street - Le Pen, Muslims, and Europe', 2 May 2002.

- *The Globalist*. 'Teenagers: The Globe's Future', 29 June 2002.

- The World Bank. *Globalization, Growth and Poverty*, The World Bank, Washington D.C., 2002.

- The World Bank. *The Economic Consequences of HIV in Russia*, The World Bank, Russia Office, 10 November 2002.

- The World Bank. *Unlocking the Employment Potential in the Middle East and North Africa. Toward a New Social Contract*, MENA Development Report, The World Bank, Washington, D.C., 2004.

- Tian, Xiang Yue et al. 'Surface Modelling of Human Population Distribution in China', *Ecological Modelling*, no. 181, 2005.

- Turner, Mark. 'The Shadow of AIDS is Casting a Pall of Darkness over the Heart of Africa', *Financial Times*, 1 July 2000.

- United Nations. *Replacement Migration: Is it a Solution to Declining and Ageing Populations?*, Population Division, Department of Economic and Social Affairs, 21 March 2000.

- United Nations. *Replacement Migration*, Shorter Revised Draft, Population Division, Department of Economic and Social Affairs, 20 December 2000.

- United Nations. *World Population Ageing: 1950-2050*, Population Division, Department of Economic and Social Affairs, 2002.

Bibliography

- United Nations. *World Population Prospects: The 2002 Revision. Highlights*, Population Division, Department of Economic and Social Affairs, February 2002.

- United Nations. *Levels and Trends of International Migration to Selected Countries in Asia*, Population Division, Department of Economic and Social Affairs, 2003.

- United Nations. *Community Realities & Responses to HIV/AIDS in Sub-Saharan Africa*, June 2003.

- United Nations. *World Economic and Social Survey: International Migration*, Population Division, Department of Economic and Social Affairs, 2004.

- United Nations. *World Population Prospects: The 2004 Revision*, United Nations, New York, 2005.

- United Nations. *Trends in Total Migrant Stock: The 2005 Revision*, Population Division, Department of Economic and Social Affairs, February 2006.

- United Nations Environment Programme. *Global Environment Outlook Year Book: an Overview of our Changing Environment 2004/5*, 2005.

- United Nations High Commissioner for Refugees. *2004 Global Refugee Trends*, Population and Geographical Data Section, Division of Operational Support, United Nations High Commissariat for Refugees, 17 June 2005.

- United Nations Programme on HIV/AIDS. *HIV/AIDS. Country Profile: India*, Population Division, Department of Economic and Social Affairs, March 2003.

- United Nations Programme on HIV/AIDS. *2004 Report on the Global HIV/AIDS Epidemic: 4th Global Report*, UNAIDS, Geneva, 2004.

- United Nations Programme on HIV/AIDS. *AIDS in Africa: Three Scenarios to 2025*, UNAIDS, Geneva, 2005.

- United Nations Programme on HIV/AIDS & The World Health Organization. *Aids Epidemic Update*, Special Report on HIV Prevention, December 2005.

- United States Census Bureau. *U.S. Interim Projections by Age, Sex, Race, and Hispanic Origin*, Washington D.C., 18 March 2004.

- Wallander, Celeste A. *The Impending AIDS Crisis in Russia: The Shape of the Problem and Possible Solutions*, Conference at the CSIS HIV/AIDS Task Force, Center for Strategic and International Studies, 14 April 2004.

- Wolf, Martin. 'The Low Cost of Better Health', *Financial Times*, 9 January 2002.

- World Health Organization. *Emerging Issues in Water and Infectious Diseases*, Geneva, 2003.

- World Health Organization. *Millennium Ecosystem Assessment. Ecosystems and Human Well-Being: A Health Synthesis*, Geneva, 2005.

2. ECONOMY

- Archibugi, Daniele & Coco, Alberto. 'Is Europe becoming the most dynamic knowledge economy in the world?', *Journal of Common Market Studies*, vol. 43, no. 3, September 2005.
- Asian Development Bank. *Asian Development Outlook 2006: Routes for Asia's Trade*, 2006.
- Asuncion-Mund, Jennifer. *India rising: A medium-term perspective*, Deutsche Bank Research, Frankfurt, May 2005.
- *L'Atlas du Monde diplomatique*. 'Mondialisation, gagnants et perdants. Derrière les mythes du libre-échange', 2005.
- *L'Atlas du Monde diplomatique*. 'Une nouvelle géopolitique. Afrique, miroir du monde', 2005.
- Aykut, Dilek & Ratha, Dilip. 'South-South FDI flows: how big are they?', *Transnational Corporations*, vol.13, no. 1, April 2004.
- Beck, Roland & Schularick, Moritz. *Russia 2010: Scenarios for Economic Development*, Deutsche Bank Research, Frankfurt, March 2003.
- Bergheim, Stefan. *Global Growth Centres 2020*, Current Issues, Deutsche Bank Research, Frankfurt, 23 March 2005.
- Bergheim, Stefan. *Global Growth Centres 2020: A Guide for Long-Term Analysis and Forecasting*, American Institute for Contemporary German Studies, Washington D.C., 23 June 2005.
- Bergheim, Stefan. *Global Growth Centres: Human Capital is the Key to Growth: Success Stories and Policies for 2020*, Deutsche Bank Research, Frankfurt, 1 August 2005.
- Bhalla, G. S., Hazell, Peter & Kerr, John. « Les perspectives de l'offre et de la demande de céréales en Inde à l'horizon 2020 », *Vision 2020 pour l'alimentation, l'agriculture et l'environnement*, International Food Policy Research Institute (IFPRI), November 1999.
- Borg, Scott. 'Economically Complex Cyberattacks', *IEEE Security and Privacy*, vol. 3, no. 6, IEEE Computer Society, 2005.
- Bräuninger, Dieter. *More Jobs for Older Workers – Against Unemployment and Early Retirement*, Current Issues, Deutsche Bank Research, Frankfurt, 14 October 2005.
- Brooks, Robin. *Population Ageing and Global Capital Flows in a Parallel Universe*, IMF Staff Papers, vol. 50, no. 2, IMF, 2003.
- Bruyas, Pierre. 'Socio-economic Aspects of the Maghreb', *Statistics in Focus*, Eurostat, April 2005.
- Centre d'études prospectives et d'informations internationales (CEPII). *European industry's place in the international division of labour: situation and prospects*, Report prepared for the Directorate-General for Trade of the European Commission, July 2004.
- Chauvin, Sophie & Lemoine, Françoise. *L'économie indienne: changements structurels et perspectives à long terme*, CEPII Working Paper no. 2005-04, April 2005.
- Christidis, Panayotis, Hernandez, Hector & Lievonen, Jorma (eds.). *Impact of Technological and Structural Change on Employment: Prospective Analysis 2020, Background Report*, Study for the Committee on Employment and Social Affairs of the European Parliament, IPTS/ESTO, March 2002.

- Citrin, Daniel & Wolfson, Alexander. 'Japan's Back', *Finance and Development*, vol. 43, no. 2, June 2006.

- Clark, Greg. *Emerging Local Economic Development Lessons from Cities in the Developed World, and Their Applicability to Cities in Developing and Transitioning Countries*, 'World Bank Urban Forum: Tools, Nuts, and Bolts', Washington, D.C., April 2002.

- Clemens Jnr, Walter C. 'Alternative Futures AD 2000-2025', *The OECD Observer*, no. 221/222, Summer 2000.

- Congressional Budget Office. *CBO's Current Economic Projections*, The Congress of the United States, August 2005.

- Congressional Budget Office. *The Budget and Economic Outlook: An Update*, The Congress of the United States, August 2005.

- Congressional Budget Office. *Global Population Ageing in the 21st Century and Its Economic Implications*, The Congress of the United States, December 2005.

- Congressional Budget Office. *The Long-Term Budget Outlook*, The Congress of the United States, December 2005.

- Crawford, Jo-Ann & Fiorentino, Roberto V. *Changing the Landscape of Regional Trading Agreements*, World Trade Organization Discussion Paper no. 8, 2005.

- Dasgupta, Susmita, Wang, Hua & Wheeler, David. *Surviving Success: Policy Reform and the Future of Industrial Pollution in China*, China's National Environmental Protection Agency and The World Bank's Country Department, March 1997.

- Dosi, Giovanni, Llerena, Patrick & Sylos Labini, Mauro. *Evaluating and Comparing the Innovation Performance of the United States and the European Union*, Expert report prepared for the TrendChart Policy Workshop 2005, 29 June 2005.

- Dussel Peters, Enrique & Dong Xue, Liu. 'Economic Opportunities and Challenges posed by China for Mexico and Central America', *DIE Studies*, no. 8, Deutsche Institut für Entwicklungspolitik, Bonn, 2005.

- Economic Commission for Africa. *Meeting the Challenges of Unemployment and Poverty in Africa*, Economic Report on Africa 2005, November 2005.

- Economist Intelligence Unit. *Foresight 2020: Economic, Industry and Corporate Trends*, March 2006.

- Eichengreen, Barry. 'The Blind Men and the Elephant', *Issues in Economic Policy* no. 1, The Brookings Institution, January 2006.

- Elmeskov, Jørgen & Scarpetta, Stefano. *New Sources of Economic Growth in Europe?*, 28th Economics Conference 2000: 'The New Millennium – Time for a new economic paradigm?', Oesterreichische Nationalbank, Vienna, 15-16 June 2000.

- European Commission. *Creating an Innovative Europe*, Report of the Independent Expert Group on R&D and Innovation appointed following the Hampton Court Summit, January 2006.

- European Commission. *EU Sectoral Competitiveness Indicators*, Enterprise and Industry Directorate-General, 'Competitiveness and economic reforms', 2005.

- European Commission. *Towards a European Research Area: Science, Technology and Innovation: Key Indicators 2005*, 2005.

- Exxon Mobil, *A Report on Energy Trends, Greenhouse Gas Emissions and Alternative Energy*, February 2004.

- Fay, Marianne & Morrison, Mary. *Infrastructure in Latin America & the Caribbean: Recent Developments and Key Challenges*, The World Bank, Washington D.C., 2005.

- Federal Reserve Board. 'Major Foreign Holders of Treasury Securities', US Department of the Treasury, January 2006.

- Fontagné, Lionel & Lorenzi, Jean-Hervé. 'Désindustrialisation, délocalisations', *Rapport du Conseil d'Analyse Economique*, (55), 2005.

- Gaulier, Guillaume, Lemoine, Françoise & Ünal-Kesenci, Deniz. *China's Emergence and the Reorganisation of Trade Flows in Asia*, CEPII, 2006-05.

- Geithner, Timothy F. *Policy Implications of Global Imbalances*, Speech at Chatham House, 23 January 2006.

- Gelauff, Georges M. & Lejour, Arjan M. *The New Lisbon Strategy: An estimation of the economic impact of reaching five Lisbon Targets*, Industrial Policy and Economic Reforms Papers no. 1, European Commission, January 2006.

- Gesellschaft für Technische Zusammenarbeit (GTZ). *Innovation Policy Trends in Selected Asian, MENA, Eastern and Western European Countries*, Eschborn, July 2005.

- Gist, John R. & Verma, Satyendra. *Entitlement Spending and the Economy: Past Trends and Future Projections*, AARP Public Policy Institute, 12 September 2002.

- Glaessner, Thomas C., Kellermann, Tom & McNevin Valerie. *Electronic Safety and Soundness: Securing Finance in a New Age*, World Bank Working Paper no. 26, The World Bank, Washington D.C., February 2004.

- Hara, Yonosuke (Chairman). *Asian Dynamism and Prospects for Technical Cooperation Policies*, Interim Report, Ministry of Economy, Trade and Industry's Study Group on Technical Cooperation in Economics, Trade and Industry, Japan, March 2004.

- Herring, Richard, Johnson, Suzanne N. & Litan Robert E. *The Top Ten Financial Risks to the Global Economy*, Conference Summary, Global Markets Institute at Goldman Sachs, September 2005.

- Heymann, Eric. *Dynamic sectors give global growth centres the edge*, Deutsche Bank Research, Frankfurt, 31 October 2005.

- Holz Carsten A. *China's Economic Growth 1978-2025: What We Know Today about China's Economic Growth Tomorrow*, Working Paper no. 8, Hong Kong University of Science & Technology, Centre on China's Transnational Relations, 3 July 2005.

- Hunt, Benjamin. *Oil Price Shocks: Can They Account for the Stagflation in the 1970s?*, IMF Working Paper, IMF, November 2005.

- International Energy Agency. *Analysis of the Impact of High Oil Prices on the Global Economy*, May 2004.

- International Monetary Fund. *2005 World Economic Outlook: Building Institutions*, September 2005.

Bibliography

- International Monetary Fund. *World Economic Outlook: Advancing Structural Reforms*, A Survey by the Staff of the IMF, April 2004.

- Italian Ministry of Economy and Finance. *Sixty Years After Bretton Woods: Developing a Vision for the Future*, Rome, 22-23 July 2004.

- Izraelewicz, Erik. *Quand la Chine change le monde* (Éditions Grasset, Paris, 2005).

- Just, Tobias. *Adapting to Demographic Trends: Major and Minor Challenges*, Deutsche Bank Research, Frankfurt, 7 July 2005.

- Kan, Suyin & Hubbard, Claire V.M. (eds.). *Latin America Consensus Forecasts* (Consensus Economics Inc., 2005).

- Karoly Lynn A. & Panis, Constantijn W.A. *The 21st Century at Work: Forces Shaping the Future Workforce and Workplace in the United States*, Prepared for the US Department of Labor, RAND Labor and Population, 2004.

- Levy, Frank & Murnane, Richard J. *How Computerized Work and Globalization Shape Human Skill Demands*, MIT, September 2005.

- Mistral, Jacques & Salzman Bernard. 'La préférence américaine pour l'inégalité', *En Temps Réel*, Cahier no. 25, 2006.

- Murray, Justin & Labonte, Marc. *Foreign Holdings of Federal Debt*, Congressional Research Service (CRS) Report for Congress, 23 November 2005.

- National Intelligence Council. 2020 Project: *'Africa in 2020', Summary of Discussion by Select Panel of US Experts*, 9 January 2004.

- National Intelligence Council. 2020 Project: *Mapping Sub-Saharan Africa's Future*, March 2005.

- Neuhaus, Marco. *Opening economies succeed. More trade boosts growth*, Current Issues, Deutsche Bank Research, Frankfurt, 11 November 2005.

- Oliveira Martins, Joaquim, Gonand, Frédéric, Antolin, Pablo, de la Maisonneuve, Christine, & Yoo, Kwang-Yeol. *The Impact of Ageing on Demand, Factor Markets and Growth*, Economics Working Paper no. 420, OECD, 29 March 2005.

- Organisation for Economic Co-operation and Development. *Sustainable Development: Critical Issues*, Policy Brief, September 2001.

- Organisation for Economic Co-operation and Development. *Economic Survey of the US*, Policy Brief, 2005.

- Organisation for Economic Co-operation and Development. *Etude Economique de la Chine, 2005*, 2005.

- Organisation for Economic Co-operation and Development. *Science, Technology and Industry Scoreboard*, 2005.

- Organisation for Economic Co-operation and Development. *Trade and Structural Adjustment: Embracing Globalisation*, 2005.

- Organisation for Economic Co-operation and Development. *African Economic Outlook 2005/2006*, 2006.

- Organisation for Economic Co-operation and Development. *Economic Survey of Japan*, Policy Brief, 2006.

- Pei, Minxin. 'The Dark Side of China's Rise', *Foreign Policy*, March/April 2006.

- Perlitz, Uwe. *Chemical Industry in China: Overtaking the Competition*, Deutsche Bank Research, Frankfurt, 25 October 2005.

- Piketty, Thomas & Saez, Emmanuel. 'The evolution of top incomes: a historical and international perspective', *American Economic Review*, vol. 6, no. 2, May 2006.

- Pitigala, Nihal. *What Does Regional Trade in South Asia Reveal about Future Trade Integration? Some Empirical Evidence*, World Bank Policy Research Working Paper, The World Bank, February 2005.

- Planning Commission, Government of India. *Report of the Committee on India Vision 2020*, New Delhi, 2002.

- Porter, Michael E. & van Opstal, Debra. 'U.S. Competitiveness 2001: Strengths, Vulnerabilities and Long-Term Priorities' in *World Economic Situation and Prospects 2006*, United Nations, New York, 2001.

- *Pravda*. 'Russia's economy to develop steadily during the current decade, German experts say', UK Pravda Interview of Mr. Norbert Walter, Chief Economist of Deutsche Bank Group, UK, 19 September 2005.

- Reinaud, Julia. *Emissions Trading and its Possible Impact on Investment Decisions in the Power Sector*, IEA Information Paper, International Energy Agency, 2003.

- Richards, Alan. 'The Global Financial Crisis and Economic Reform in the Middle East', *Middle East Policy Council Journal*, vol. VI, no. 3, February 1999.

- Richardson, Pete (ed.). *Globalisation and Linkages: Macro-Structural Challenges and Opportunities*, Working Paper no.181, OECD Economics Department, Paris, 1997.

- Roland-Holst, David, Verbiest, Jean-Pierre & Zhai, Fan. *Growth and Trade Horizons for Asia: Long-Term Forecasts for Regional Integration*, ERD Working Paper no. 74, Asian Development Bank, November 2005.

- Ryan P. *Global Competitiveness: U.S. Firms Emerge*, Global Strategic Research Team, Marubeni Research Institute, August 2002.

- Scapolo, Fabiana, Geyer, Anton, Boden, Mark, Döry, Tibor & Ducatel, Ken. *The Future of Manufacturing in Europe 2015-2020: The Challenge for Sustainability*, Institute for Prospective Technological Studies (IPTS), European Commission Joint Research Centre, March 2003.

- Schaffer, Teresita C. & Mitra, Pramit. *India as a Global Power?*, Deutsche Bank Report, 16 December 2005.

- Shukla, Priyadarshi. *Economic Environment Modeling: Policy Insights for India*, Economic and Environmental Modeling Workshop, New Delhi, 19-20 January, 2004.

- Strecker Downs, Erica. *China's Quest for Energy Security* (RAND Corporation, Santa Monica, 2000).

- The World Bank. *Finance for Growth: Policy Choices in a Volatile World*, Working Paper 22239, A World Bank Policy Research Report, The World Bank and Oxford University Press, April 2001.

- The World Bank. *Trade, Investment, and Development in the Middle East and North Africa. Engaging with the World*, MENA Development Report, Washington D.C., August 2003.

- The World Bank. *Middle East and North Africa. Oil Booms and Revenue Management: Economic Developments and Prospects 2005*, Washington D.C., 2005.

- The World Bank. *Unlocking the Employment Potential in the Middle East and North Africa - Towards a new Social Contract*, MENA Development Report, Washington D.C., 2005.

- The World Bank. *A Time to Choose: Caribbean Development in the 21st Century*, Report no. 31725, Caribbean Country Management Unit, Poverty Reduction and Economic Management Unit, Latin America and the Caribbean Region, 7 April 2005.

- The World Bank. *Global Development Finance: The Development Potential of Surging Capital Flows*, Washington D.C., 2006.

- The World Bank. *Global Economic Prospects: Economic Implications of Remittances and Migration*, Washington D.C., 2006.

- The World Bank. *Where is the Wealth of Nations? Measuring Capital for the 21st Century*, The International Bank for Reconstruction and Development and The World Bank, Washington D.C., 2006.

- The World Bank. *World Development Report: Equity and Development*, Washington D.C., 2006.

- Toshida, Seiichi & Nakamura, Yoichi. *Long-Term Economic Forecast of the Japanese Economy (2001-2025)*, Japan Center for Economic Research, March 2001.

- UK Department of Trade and Industry. *Trade and Investment Implications of EU Enlargement*, Europe and World Trade Directorate, April 2004.

- UK Department of Trade and Industry. *Creating a Low Carbon Economy*, Second Annual Report on the Implementation of the Energy White Paper, July 2005.

- United Nations Conference on Trade and Development. *Enhancing the Contribution to Development of the Indigenous Private Sector in Africa: Challenges and Opportunities for Asia-Africa Cooperation*, Strategic framework for the AFRASIA Business Council, Investment and Enterprise Competitiveness Branch and the AFRASIA Business Council Executive Support Group, 2003.

- United Nations, Economic Commission for Africa. *Economic Report on Africa: Meeting the Challenges of Unemployment and Poverty in Africa*, 2005.

- United Nations, UNCTAD. *Economic Development in Africa: Rethinking the Role of Foreign Direct Investment*, 2005.

- United Nations, UNCTAD. *World Investment Report: The Shift Towards Services*, 2004.

- United Nations, UNCTAD. *World Investment Report: Transnational Corporations and the Internationalization of R&D*, 2005.

- United Nations. *World Economic Situation and Prospects 2006*, Executive Summary, 2006.

- US Department of the Treasury. *The Debt to the Penny and Who Holds It*, Bureau of the Public Debt, last updated 24 March 2006.

- Walter, Norbert. *Globalisation: The World Economy in the Year 2025*, Deutsche Bank Group, 24 June 2005.

❚ Weyant, John. 'India, Sustainable Development, and The Global Commons', in Janardhan Rao, N. (ed.) *India: An Emerging Economic Powerhouse*, The ICFAI University Press, April 2005.

❚ Williamson J. 'What Follows the USA as the World's Growth Engine?', India Policy Forum Public Lecture, 2005.

❚ Wilson, Dominic & Purushothaman, Roopa. *Dreaming with BRICs: The Path to 2050*, Global Economics Paper no. 99, Goldman Sachs, 1 October 2003.

❚ Wilson, Dominic, Purushothaman, Roopa & Fiotakis, Themistoklis. *The BRICs and Global Markets: Crude, Cars and Capital*, Global Economics Paper no. 118, Goldman Sachs, 14 October 2004.

❚ World Economic Forum. *The Global Competitiveness Report 2005-2006*, New York, 2005.

❚ Zedillo, Ernesto, Messerlin, Patrick & Nielson Julia. *Trade for Development*, UN Millennium Project, Task Force on Trade, 2005.

❚ Zhihong, Wei. *China's Challenging Fast Track*, International Atomic Energy Agency (IAEA) Bulletin 46/1, June 2004.

3. ENERGY

❚ Aden, Nathaniel, Sinton, Jonathan, Stern, Rachel & Levine, Mark. *Evaluation of China's Energy Strategy Options*, Report prepared for and with the support of the China Sustainable Energy Program, 16 May 2005.

❚ Alameddine, Chirine H. *Le Développement urbain au Moyen-Orient et en Afrique du Nord*, Note sectorielle, The World Bank, August 2005.

❚ Arctic Climate Impact Assessment (ACIA). *Impacts of a Warming Arctic* (Cambridge University Press, 2004).

❚ Auer, Josef. *Energy Prospects after the Petroleum Age*, Deutsche Bank Research, 2 December 2004.

❚ Baily, Martin Neil & Kirkegaard, Jacob F. *The US Economic Outlook*, Institute For International Economics, Washington D.C., 15 September 2004.

❚ Baker, Murl et al. *Conflict Timber: Dimensions of the Problem in Asia and Africa*, Final Report Submitted to the United States Agency for International Development, ARD, 2003.

❚ Bhalla, G.S. et al. *Prospects for India's Cereal Supply and Demand to 2020*, Discussion Paper 29, International Food Policy Research Institute, Washington D.C., November 1999.

❚ British Petroleum. *BP Statistical Review of World Energy 2002*, London, June 2002.

❚ British Petroleum. *Putting Energy in the Spotlight: BP Statistical Review of World Energy 2005*, London, June 2005.

❚ British Petroleum. *Quantifying Energy: BP Statistical Review of World Energy 2006*, London, June 2006.

Bibliography

- Cambridge Energy Researches Associates. *Dawn of a New Age: Global Energy Scenarios*, CERA Conference, Cambridge, Massachusetts, 6 December 2005.

- Chellaney, Brahma. 'India's Future Security Challenge: Energy Security', in *India as a New Global Leader*, The Foreign Policy Centre, 2005.

- Cleuntinx, Christian. *The EU-Russia Energy Dialogue*, DG for Energy and Transport, European Commission, Vienna, December 2003..

- *Climate Change, Energy and Sustainable Development: How to Tame King Coal?*, Vision Paper, Coal Working Group, Le Délégué Interministeriel au Développement Durable, France, 9 June 2005. Revised version, 12 January 2006.

- Committee on the Science of Climate Change. *Climate Change Science* (National Academy Press, Washington D.C., 2001).

- Congressional Research Service (CRS). *Caspian Oil and Gas: Production and Prospects*, 4 March 2005.

- Congressional Research Service (CRS). *Rising Energy Competition and Energy Security in Northeast Asia: Issues for US Policy*, 14 July 2004.

- Connors, Stephen R. & Schenler, Warren W. *Climate Change and Competition – On a Collision Course*, Proceedings of the 60th American Power Conference, Chicago, 14-16 April 1998.

- Deffeyes, Kenneth S. *Hubbert's Peak: The Impending World Oil Shortage* (Princeton University Press, NJ, 2001).

- Delgado, Christopher et al. *Livestock to 2020 – The Next Food Revolution* (ILRI, Addis-Ababa, Ethiopia, 1999).

- Deutch, John M. *Future United States Energy Security Concerns*, Report No. 115, MIT Joint Program on the Science and Policy of Global Change, September 2004.

- Deutch, Philip J. 'Energy Independence', *Foreign Policy*, November/December 2005.

- Diamond, Rick. 'A lifestyle-based scenario for US buildings: Implications for energy use', *Energy Policy*, vol. 31, 2003.

- Energy Information Administration. *International Energy Outlook 2005*, US Department of Energy, July 2005.

- Energy Information Administration. *Petroleum Supply Annual 2004*, vol. 1, Office of Oil and Gas, US Department of Energy, June 2005.

- Energy Information Administration. *Annual Energy Outlook 2006*, US Department of Energy, February 2006.

- Energy Information Administration. *International Energy Outlook 2006*, US Department of Energy, June 2006.

- EURACTIV. *Géopolitique des approvisionnements énergétiques de l'UE*, 20 July 2005.

- EUROGULF: *An EU-GCC Dialogue for Energy Stability and Sustainability*, Final Research Report (as presented at the Concluding Conference in Kuwait, 2-3 April 2005), Project Ref.: 4.1041/D/02-008-S07 21089, 2005.

- European Commission. *European Union Energy Outlook 2020, Energy in Europe: Special Issue*, November 1999.
- European Commission. *Towards a European Strategy for the Security of Energy Supply*, Green Paper (COM(2000) 769 final), 29 November 2000.
- European Commission. *European Energy and Transport Trends to 2030*, DG for Energy and Transport, January 2003.
- European Commission. *Doing More With Less*, Green Paper on Energy Efficiency (COM 2005) 265 final), Brussels, 22 June 2005.
- European Commission. *A European Strategy for Sustainable, Competitive and Secure Energy*, Green Paper, COM(2006) 105 final, {SEC(2006) 317}, Brussels, 8 March 2006.
- European Commission. *A European Strategy for Sustainable, Competitive and Secure Energy: What is at Stake*, Background document, Commission staff working document, Annex to the Green Paper, {COM(2006) 105 final}, SEC(2006) 317/2, Brussels, 2006.
- European Council for Automotive R&D (EUCAR), Conservation of Clean Air and Water in Europe (CONCAWE) and Joint Research Centre of the European Commission. *Well-to-Wheels analysis of future automotive fuels and powertrains in the European context*, Well-to-Wheels Report, Version 1b, January 2004.
- European Environment Agency. *Energy and Environment in the European Union*, Environmental Issue Report no. 31, May 2002.
- European Environment Agency. *Impact of Europe's Changing Climate*, EEA Report no. 2/2004, August 2004.
- European Environment Agency. *Sustainable Use and Management of Natural Resources*, EEA Report no. 9/2005, January 2005.
- European Environment Agency. *How Much Bioenergy Can Europe Produce Without Harming the Environment?*, EEA Report no. 7/2006, June 2006.
- European Environment Agency. *Using the Market for Cost-Effective Environmental Policy*, EEA Report no. 1/2006, January 2006.
- European Policy Centre. *Energy Security: A European Perspective*, 18 February 2005.
- Eurostat. *Energy: Yearly Statistics. Data 2004*, June 2006
- Exxon Mobil. *Future Energy Trends and Development*, February 2004.
- Exxon Mobil. *Greenhouse Gas Emissions and Alternative Energy*, February 2004.
- Exxon Mobil. *The Outlook for Energy. A 2030 View*, 2005.
- Food and Agriculture Organization. *World Resources 2005*, 2005.
- Fourth World Water Forum. *African Regional Document: Water Resources Development in Africa*, Mexico 16-22 March 2006.
- Government of Canada. *Action on Climate Change: Considerations for an Effective International Approach*, Discussion Paper for the Preparatory Meeting of Ministers for Montreal 2005: United Nations Climate Change Conference, 2005.

- Guterl, Fred. 'Another Nuclear Dawn', *Newsweek*, 6 February 2006.

- Hill, Fiona. *Energy Empire: Oil, Gas and Russia's Revival*, The Foreign Policy Centre, London, September 2004.

- Intergovernmental Panel on Climate Change (IPCC). *Climate Change 2001, Synthesis Report*, 2001.

- International Atomic Energy Agency. *Energy, Electricity and Nuclear Power Estimates for the Period up to 2030*, Vienna, July 2005.

- International Energy Agency. *India - A Growing International Oil and Gas Player*, IEA, Paris, 2000.

- International Energy Agency. *Curbing Energy Demand in India? Where to Start?* Findings from an IEA Feasibility Study, IEA, Paris, 2002.

- International Energy Agency. *Energy for 2050: Scenarios for a Sustainable Future*, IEA/OECD, Paris, 2003.

- International Energy Agency. *World Energy Investment Outlook 2003*, 2003.

- International Energy Agency. *Key World Energy Statistics*, 2004.

- International Energy Agency. *World Energy Outlook 2004*, 2004.

- International Energy Agency. *Findings of Recent IEA Work*, 2005.

- International Energy Agency. *World Energy Outlook 2005: Middle East and North Africa Insights*, 2005.

- International Energy Agency/OPEC workshop. *Oil Outlook and Investment Prospects*, 2004.

- Khatib, Hisham. *Energy Considerations- Global Warming Perspectives*, Energy Permanent Monitoring Panel of the World Federation of Science, 19 August 2004.

- Kim, Marina. 'Russian Oil and Gas: Impacts on Global Supplies to 2020', *Australian Commodities*, vol. 12, no.2, June Quarter 2005.

- Larson, Alan. 'Geopolitics of Oil and Natural Gas', *Economic Perspective*, vol. 9, no. 2, May 2004.

- Lee, Julian. *Future Russian Oil Production: The CGES View*, Centre for Global Energy Studies, CERI 2004 Oil Conference, 28-30 March 2004.

- Le-Huu, Ti. 'Natural Disasters: overview of recent trends in natural disasters in Asia and the Pacific', *Water Resources*, December 2005.

- Leijonhielm, Jan & Larsson, Robert. *Russia's Strategic Commodities: Energy and Metals as Security Levers*, Swedish Defence Research Agency, November 2004.

- Luft, Gal. *Fueling the dragon: China's race into the oil market*, Institute for the Analysis of Global Security (IAGS), Washington D.C.

- Lynch, Dov (ed.). *The South Caucasus: a Challenge for the EU*, Chaillot Paper no.65, EUISS, Paris, December 2003.

- Massachusetts Institute of Technology. *The Future of Nuclear Power*, MIT Press, 2003.

- Milov, Vladimir (ed.). *Deepening the integration between energy producing and consuming nations*, Institute of Energy Policy (Moscow), 25 January 2006.

- Muttit, Greg. *Crude Designs. The Rip-Off of Iraq's Oil Wealth*, Platform, 2005.
- Organization of the Petroleum Exporting Countries. *Annual Statistical Bulletin*, 2004.
- Organization of the Petroleum Exporting Countries. *Oil Outlook to 2025*, OPEC Secretariat Paper, 10th International Energy Forum, Doha, 22-24 April 2006.
- Paillard, Christophe-Alexandre. 'L'influence des prix du pétrole sur l'économie mondiale', *Questions internationales*, no. 18, La Documentation française, Paris, March/April 2006.
- Paillard, Christophe-Alexandre. *Strategies for Energy: Which Way Forward for Europe?*, Fondation Robert Schuman, Paris, February 2006.
- Philibert, Cédric. *The Present and Future Use of Solar Thermal Energy*, International Energy Agency, The InterAcademy Council, 2005.
- RAND Corporation. *E-vision 2000: Key Issues That Will Shape our Energy Future*, Science and Technology Policy Institute, Rand, Santa Monica, June 2001.
- Rempel. H. *Will the Hydrocarbon Era Finish Soon?*, Federal Institute for Geosciences and Natural Resources, Hannover, 23 May 2000.
- Royal Commission on Environmental Pollution. *Biomass as a Renewable Energy Source*, United Kingdom, May 2004.
- Russ, Peter, Ciscar, Juan Carlos & Szabó Laszlo. *Analysis of Post-2012. Climate Policy Scenarios with Limited Participation*, Institute for Prospective and Technological Studies (IPTS), 2005.
- Russian Academy of Science. *Review of Long-term Energy Scenarios: Implications for Nuclear Energy*, Moscow, 25-26 April 2002.
- Saghir, Jamal. *Energy and Poverty: Myths, Links and Policy Issues*, Energy Working Notes, Energy and Mining Sector Board, The World Bank, May 2005.
- Simmons, Matthew R. *Twilight in the Desert: The Coming Saudi Oil Shock and the World Economy* (John Wiley & Sons Inc., New Jersey, 2005).
- State of the Union Address by President George W. Bush, United States Capitol, Washington D.C., 2006.
- The Brookings Institution. *Energy Security: Responding to the Challenge*, Brookings Briefing, Washington, 29 November 2005.
- The World Bank. *Middle East and North Africa Regional Water Initiative*, Regional Water Initiative Report, Spring 2002.
- 'The Year Ahead', *Oil & Gas Journal*, 2 January 2006.
- Titus, James G. & Narayanan, Vinjay. *The Probability of Sea Level Rise*, US Environmental Protection Agency, Washington D.C., 1995.
- True, Warren. 'WGC: Growth in gas trade faces range of challenges', *Oil and Gas Journal*, 12 June 2006.
- UK Department of Trade and Industry. *Our Energy Future - Creating a Low Carbon Economy*, UK Energy White Paper, DTI, March 2003.

- UK Department of Trade and Industry. *The Energy Challenge. Energy Review Report 2006*, DTI, July 2006.

- United Nations. *World Urbanization Prospects: The 2003 Revision*, Department of Economic and Social Affairs, Population Division, 2003.

- United Nations. *The Energy Challenge for Achieving the Millennium Development Goals*, Department of Economic and Social Affairs and UN-Energy, 2005.

- United Nations Environment Programme. '2032: Choices for the Future', *Global Environment Outlook 3*, 2002.

- United Nations Environment Programme. *The Asian Brown Cloud: Climate and Other Environmental Impacts*, Center for Clouds, Chemistry and Climate, 2002.

- von Braun, Joachim et al. *New Risks and Opportunities for Food Security. Scenario Analyses for 2015 and 2050*, Discussion Paper 39, International Food Policy Research Institute, Washington D.C., 2005.

- Willenborg, Robert et al. *Europe's Oil Defences. An Analysis of Europe's Oil Supply Vulnerability and its Emergency Oil Stockholding Systems*, The Clingendael Institute, The Hague, 2004.

- Wiser, Ryan & Bolinger, Mark. *An Overview of Alternative Fossil Fuel Price and Carbon Regulation Scenarios*, Environmental Energy Technologies Division, Lawrence Berkeley National Laboratory, October 2004.

- World Energy Council. *Reflections on Energy and Climate Change*, Working Paper, July 2004.

- World Energy Council. *The World Energy Book*, 2005.

- Yergin, Daniel. 'Questions of Oil', *The Economist: The World in 2006*, 2005.

- Zittel, Werner & Schindler, Jörg. 'The Countdown for the Peak of Oil Production Has Begun – But What are the Views of the Most Important International Energy Agencies?', *EnergyBulletin.net/EnergieKrise.de*, 14 October 2004.

4. ENVIRONMENT

- African Development Bank et al. *Poverty and Climate Change: Reducing the Vulnerability of the Poor through Adaptation*. A Contribution to the Eighth Conference of the Parties to the United Nations Framework Convention on Climate Change, October 2002.

- Alameddine, Chirine. H. *Le Développement urbain au Moyen-Orient et en Afrique du Nord*, Note sectorielle, The World Bank, August 2005.

- Andersen, Stephen O. et al. (eds.). *Special Report on Safeguarding the Ozone Layer and the Global Climate System: Issues Related to Hydrofluorocarbons and Perfluorocarbons*, (Cambridge University Press, April 2005).

- Arctic Climate Impact Assessment (ARCIA). *Impacts of a Warming Arctic* (Cambridge University Press, 2004).

- *Asia-Pacific Regional Document*, 4th World Water Forum, Mexico, 16-22 March 2006.
- Asian Development Bank. *Handbook on Environment Statistics, Development Indicators and Policy Research Division*, Economics and Research Department, Asian Development Bank, April 2002.
- Baker, Murl et al. *Conflict Timber: Dimensions of the Problem in Asia and Africa*, Final Report Submitted to the United States Agency for International Development, ARD, 2003.
- Barreto, Leonardo & Turton, Hal. *Impact Assessment of Energy-related Policy Instruments on Climate Change and Security of Energy Supply*, Interim Report, International Institute for Applied Systems Analysis, Austria, 18 January 2005.
- Bhalla, G.S. et al. *Prospects for India's Cereal Supply and Demand to 2020*, International Food Policy Research Institute, Discussion Paper 29, Washington D.C., November 1999.
- Biermann, Frank. *Between the United States and the South: Strategic Choices for European Climate Policy*, Global Governance Working Paper, no. 17, June 2005.
- Black, Richard. *Environmental Refugees: Myth or Reality?*, Working Paper No. 34, New Issues in Refugee Research, University of Sussex, March 2001.
- Blanchard, Odile. *The Bush Administration Climate Proposal: Rhetoric and Reality*, CFE Policy Paper, Institut français des relations internationales (IFRI), March 2003.
- Boberg, Jill. *Liquid Assets: How Demographic Changes and Water Management Policies Affect Freshwater Resources* (MG-358, Rand Corporation, Santa Monica, 2005).
- Brown, Lester R. *Plan B: Rescuing a Planet under Stress and a Civilization in Trouble* (W.W. Norton, New York, September 2003).
- Bruisma, Jelle (ed.). *World Agriculture Towards 2015/2030: a FAO Perspective* (Food and Agriculture Organization, Rome, 2005).
- Center for Health and the Global Environment (Harvard Medical School). *Climate Change Futures: Health, Ecological and Economic Dimensions*, November 2005.
- Center for Strategic and International Studies (CSIS). *The Future of Water*, CSIS Program, Issue 7, 25 February 2003.
- Chenggui, Li & Hongchun, Wang. 'China's Food Security and International Trade', *China & World Economy*, no. 4, 2002.
- 'China's Development and the Environment', Interview with Elizabeth Economy, *Harvard Asia Quarterly*, vol. VII, no.1, Winter 2003.
- Chinese Ministry of Environment. *Report on the State of the Environment in China*, 2004.
- Clay, Jason. *World Agriculture and the Environment: A Commodity-by-Commodity Guide* (Island Press, Washington D.C., March 2004).
- Committee on the Science of Climate Change. *Climate Change Science*, National Academy Press, Washington D.C., 2001.
- Connors, Stephen R. & Schenler, Warren W. *Climate Change and Competition: On a Collision Course?* (MIT, 1999).
- Cooper Ramo, Joshua. *The Beijing Consensus*, The Foreign Policy Centre, 2004.

Bibliography

- Curry, Ruth & Mauritzen, Cecilie. 'Dilution of the Northern North Atlantic Ocean in Recent Decades', *Science*, vol. 308, 2005.

- Darwin, Roy, Tsigas, Marinos, Lewandrowski, Jan & Raneses, Anton. 'Climate change, world agriculture and land use', Chapter 9 in *Global Environmental Change and Agriculture: Assessing the Impacts*, Frisvold, G. and Kuhn, B. (eds.) (Edward Elgar Publishing Ltd, Cheltenham, UK, 1999).

- Delgado, Christopher et al. *Livestock to 2020 – The Next Food Revolution* (ILRI, Addis-Ababa, Ethiopia, 1999).

- Dlugolecki, Andrew. *Climate Change and Mounting Financial Risks: What are the Options?*, Background paper for The Hague Conference on Environment, Security and Sustainable Development, 9-12 May 2004.

- Economy Elizabeth C. 'Will China Face an Environmental Meltdown?', *The Globalist*, 15 June 2004.

- Economy Elizabeth C. 'Can China go green?', *The Globalist*, 18 June 2004.

- Energy Information Administration (EIA). *China: Environmental Issues*, EIA Country Analysis Brief, US Department of Energy, July 2003.

- European Commission. *Progress Towards Achieving the Community's Kyoto Target*, {COM(2005) 655 final}, Brussels, 15 December 2005.

- European Commission. *A European Strategy for Sustainable, Competitive and Secure Energy. What is at Stake*, Background document, Annex to the Green Paper, {COM(2006) 105 final}, SEC(2006) 317/2, Brussels, 2006.

- European Environment Agency. *Energy and Environment in the European Union*, Environmental Issue Report no. 31/2002, May 2002.

- European Environment Agency. *Europe's Environment: The Third Assessment*, EEA Environmental Assessment Report no. 10/2003, May 2003.

- European Environment Agency. *Impact of Europe's Changing Climate*, EEA Report no. 2/2004, August 2004.

- European Environment Agency. *European Environment Outlook*, EEA Report no. 4/2005, September 2005.

- European Environment Agency. *The European Environment: State and Outlook 2005*, EEA State of Environment Report no.1/2005, November 2005.

- European Environment Agency. *Vulnerability and Adaptation to Climate Change in Europe*, Technical Report no. 7/2006, December 2005.

- European Environment Agency. *Sustainable Use and Management of Natural Resources*, EEA Report no. 9/2005, January 2005.

- European Environment Agency. *Using the Market for Cost-Effective Environmental Policy*, EEA Report no.1/2006, January 2006.

- Fischer, Gunther, Shah, Mahendra & van Velthuizen, Harrij. *Climate Change and Agricultural Vulnerability*, Special Report by the International Institute for Applied Systems Analysis, as a contribution to the World Summit on Sustainable Development, Johannesburg, 2002.

- Food and Agriculture Organization. *Global Forest Resources Assessment*, 2005.
- Food and Agriculture Organization. *State of the World's Forests 2001*, 2001.
- Food and Agriculture Organization. *State of the World's Forests 2003*, 2003.
- Food and Agriculture Organization. *State of the World's Forests 2005*, 2005.
- Food and Agriculture Organization. *World Agriculture: Towards 2015/2030*, July 2003.
- Ford Runge, C. 'Food Security and Globalization', *The Globalist*, January 17, 2004.
- Ford Runge, C. et al. *Ending Hunger in Our Lifetime: Food Security and Globalization* (Johns Hopkins University Press/International Food Policy Research Institute, August 2003).
- Gleick, Peter (ed.). *The World's Water 2004-2005* (Island Press, Washington D.C., November 2004).
- Glover, Linda K. & Earle, Sylvia A. (eds.), *Defying Ocean's End: An Agenda for Action* (Island Press, Washington D.C., October 2004).
- Grübler, Arnulf et al. 'Emissions Scenarios: A Final Response', *Energy & Environment*, vol. 15, no.1, 2004.
- Halsnæs, Kirsten & Shukla, Priyadarshi R. *Mainstreaming International Climate Agenda in Economic and Development Policies*, World Meteorological Organisation, 2005.
- Hatun, Hjalmar et al. 'Influence of the Atlantic Subpolar Gyre on the Thermocline circulation', *Science*, vol. 309, 2005.
- Houghton, John. *Global Warming: The Complete Briefing* (Cambridge University Press, August 2004).
- Houghton, John, Ding, Yihui, Griggs, David J., Noguer, Maria, Van der Linden, Paul J. & Xiaosu, Dai (eds.). *Climate Change 2001: The Scientific Basis*, Report of the IPCC (Cambridge University Press, UK, 2001).
- Hurell, James W., Kushnir, Yochanan, Ottersen, Geir & Visbeck, Martin. 'The North Atlantic Oscillation: Climate Significance and Environmental Impact', *Geophysical Monograph* no. 134, American Geophysical Union, 2003.
- Institute for Prospective Technological Studies (IPTS). *Analysis of Post 2012 Climate Policy Scenarios with Limited Participation*, EUR 21758 EN, 2005.
- Intergovernmental Panel on Climate Change (IPCC). *Strategies for Adaptation for Sea Level Rise*, 1991.
- Intergovernmental Panel on Climate Change (IPCC). *Climate Change 2001: Impacts, Adaptation and Vulnerability*, Report of the IPCC Panel on Climate Change, 2001.
- International Energy Agency. *Deploying Climate-friendly Technologies through Collaboration with Developing Countries*, IEA Information Paper, November 2005.
- International Energy Agency. *World Energy Outlook 2002*, 2002.
- International Energy Agency. *World Energy Outlook 2005*, 2005.

- Kent, Mary M. & Yin, Sandra. 'Controlling Infectious Diseases', *Population Bulletin*, vol. 61, no. 2, June 2006.

- Lang, Tim & Heasman, Michael. *Food Wars: The Global Battle for Mouths, Minds, and Markets* (Earthscan, London, September 2004).

- Larsen, Janet. 'Toward Global Meltdown?', *The Globalist*, 16 February 2004.

- Liu, Yingling. *Shrinking Arable Lands Jeopardizing China's Food Security*, Worldwatch Institute, 18 April 2006.

- Meadows, Donella H., Randers, Jorgen, & Meadows, Dennis L. *Limits to Growth: The 30-Year Update* (Chelsea Green, Vermont, May 2004).

- Meckling, Jonas. *Transatlantic Interdependence in US Climate Change Policy. Cross-Border State-Business Relations Challenging State Autonomy*, Global Governance Working Paper no.16, May 2005.

- Messer, Ellen, Cohen, Marc J. & D'Costa, Jashinta. *Food, Agriculture, and the Environment. Food from Peace, Breaking the Links between Conflict and Hunger*, Discussion Paper 24, International Food Policy Research Institute, June 1998.

- Metz, Bert, Davidson, Ogunlade, de Connick, Heleen, Loos, Manuela & Meyer, Leo (eds.), *IPCC Special Report on Carbon Dioxide Capture and Storage* (Cambridge University Press, September 2005).

- Metz, Bert, Davidson, Ogunlade, Swart, Rob & Pan, Jiahua (eds.), *Climate Change 2001: Mitigation*, Report of the Intergovernmental Panel on Climate Change (IPCC), (Cambridge University Press).

- Myers, Norman. *Environmental Refugees: an Emergent Security Issue*, 13th Economic Forum, Session III – Environment and Migration, Prague, 23-27 May 2005.

- Nakicenovic, Nebojsa & Swart, Rob (eds.). *Emissions Scenarios 2000*, Special Report of the Intergovernmental Panel on Climate Change (Cambridge University Press).

- Nankivell, Nathan. 'The National Security Implications of China's Emerging Water Crisis', *China Brief*, vol. 5, Issue 17, The Jamestown Foundation, 2 August 2005.

- National Research Council. *Abrupt Climate Change: Inevitable Surprises* (National Academy Press, Washington D.C., May 2002).

- Ohlsson, Leif. *Arguing the Case for an Environmental Marshall Plan*, Background Paper for The Hague conference on Environment, Security and Sustainable Development, 9-12 May 2004..

- Organisation for Economic Co-operation and Development. *OECD Environmental Outlook 2001*, 2001.

- Organisation for Economic Co-operation and Development. *Organic Agriculture: Sustainability, Markets and Policies*, May 2003.

- Organisation for Economic Co-operation and Development. *Lessons Learned in Dealing with Large-Scale Disasters*, Report of the General Secretariat Advisory Unit on Multi-disciplinary issues, 15 September 2003.

- Organisation for Economic Co-operation and Development. *Approaches for Future International Cooperation*, 14 June 2005.

- Parry, Martin, Rosenzweig, Cynthia, Iglesias, Ana, Fischer, Günther, & Livermore, Matthew. 'Climate Change and World Food Security: A New Assessment (1990-2080)', *Global Environmental Change*, vol. 9, Supplement 1, 1999.

- Pirages, Dennis & DeGeest, Theresa. *Ecological Security: An Evolutionary Perspective on Globalization* (Rowman & Littlefield, Lanham, Maryland, September 2003).

- Postel, Sandra L. 'Water for food production: will there be enough in 2025?', *Bioscience*, vol. 48, no. 8, 1998.

- Querquin, Francois et al. *World Water Actions: Making Water Flow for All* (Earthscan, London, December 2003).

- Rosegrant, Mark W., Paisner, Michael S., Meijer, Siet & Witcover, Julie. 2020 *Global Food Outlook. Trends, Alternatives, and Choices. A 2020 Vision for Food, Agriculture, and the Environment Initiative*, International Food Policy Research Institute, Washington, D.C., August 2001.

- Sarofin, Marcus S., Forest, Chris E., Reiner, David M. & Reilly, John M. *Stabilisation and Global Climate Policy*, Report no.116 of the Massachusetts Institute of Technology (MIT) Joint Program on The Science and Policy of Global Change, July 2004.

- Scherr Sara J., *Soil Degradation: A Threat to Developing-Country Food Security by 2020?* Discussion Paper 27, International Food Policy Research Institute, Washington D.C., 1999.

- Schwartz, Peter & Randall, Doug. *An Abrupt Climate Change Scenario and Its Implications for the United States' National Security*, Massachusetts Institute of Technology (MIT), October 2003.

- Shaefer, Olivier. *Integrating Renewable Energy Sources: Targets and Benefits of Large-Scale Deployment of Renewable Energy Sources*, Background Paper for The Hague conference on Environment, Security and Sustainable Development, 9-12 May 2004.

- Siebenhüner, Bernd. *The Changing Role of Nation States in International Environmental Assessments: The Case of the IPCC*, Global Governance Project Working Paper no.7, Global Governance Project, July 2003.

- Silva, Patricio. *Evaluating US States' Climate Change Initiatives*, Prepared for the Meeting: Federalism and US Climate Change Policy, Business and Policy Implications of US States' Climate Actions, CFE-IFRI, Paris, 24 May 2004.

- Speth, James Gustave. *Red Sky at Morning: America and the Crisis of the Global Environment* (Yale University Press, New Haven, March 2004).

- Strauss, Steven & Bradshaw, H.D. (eds.). 'The Bioengineered Forest: Challenges for Science and Society', *Resources for the Future*, August 2004.

- Swedish International Development Cooperation Agency. *Let it Reign: The New Water Paradigm for Global Food Security*, Stockholm International Water Institute (SIWI), 2005.

- Swedish Water House. *Investing in the Future: Water's Role in Achieving the Millennium Development Goal*, 2005.

- The International Climate Change Task Force. *Meeting the Climate Challenge*, January 2005.

- The White House. *Global Climate Change Policy Book*, United States, February 2002.

- The World Bank. *Middle East and North Africa Regional Water Initiative*, Regional Water Initiative Report, Spring 2002.

Bibliography

- The World Bank. *Environment Matters: Annual Review 2005*, Washington D.C., 2005.

- The World Bank. *Issues and Dynamics: Urban Systems in Developing East Asia* (undated).

- Ti, Le-Huu. 'Natural Disasters: overview of recent trends in natural disasters in Asia and the Pacific', *Water Resources*, Journal 217, United Nations Economic and Social Commission for Asia and the Pacific, December 2005.

- Titus, James G. & Narayanan, Vinjay. *The Probability of Sea Level Rise*, US Environmental Protection Agency, Washington D.C., 1995.

- TOTAL. *Notre énergie en partage, Rapport sociétal et environnemental 2004*, TOTAL, 2004.

- United Nations. *World Urbanization Prospects: The 2003 Revision*, Population Division, Department of Economic and Social Affairs, United Nations, 2003.

- United Nations. *Report of the Workshop on New Emissions Scenarios*, UNEP/WMO/IPCC, 1 July 2005.

- United Nations Development Programme. *World Resources 2002-2004: Decisions for the Earth: Balance, Voice and Power*, UNDP and UNEP in collaboration with The World Bank, and World Resources Institute, June 2003.

- United Nations Development Programme. *Sustainable Difference: Energy and Environment to Achieve the Millennium Development Goals*, UNDP, 2005.

- United Nations Environment Programme. *Assessing Human Vulnerability to Environmental Change. Concepts, Issues, Methods and Case Studies*, 2002.

- United Nations Environment Programme. *Global Environmental Outlook 3*, 2002.

- United Nations Environment Programme. *The Asian Brown Cloud: Climate and Other Environmental Impacts*, Center for Clouds, Chemistry and Climate`, 2002.

- United Nations Environment Programme. *Global Environment Outlook Year Book 2004-05: An Overview of Our Changing Environment*, 2005.

- United Nations Environment Programme. *Challenges to International Waters – Regional Assessments in a Global Perspective*, 2006.

- United States Department of State. *White House Climate Change Review*, Interim Report, 11 June 2001.

- United States Environmental Protection Agency. *Analysis of Costs to Abate International Ozone-Depleting Substance Substitute Emissions*, USEPA, Office of Air and Radiation, June 2004.

- Victor, David G. *Climate Change: Debating America's Policy Options – A Council Policy Initiative* (The Council on Foreign Relations Press, USA, 2004).

- von Braun, Joachim et al., *New Risks and Opportunities for Food Security: Scenario Analyses for 2015 and 2050*, Discussion Paper 39, International Food Policy Research Institute, Washington D.C., 2005.

- van der Gaag, Nicole, van Imhoff, Evert & van Wissen, Leo. 'Internal migration scenarios and regional population projections for the European Union', *International Journal of Population Geography*, vol. 6, no.1, 2000.

▌ Watson R.T. & the Core Writing Team (eds.). *IPCC Third Assessment Report: Climate Change 2001*, Synthesis Report, IPCC, Geneva, Switzerland.

▌ World Health Organization. *Heat-waves: Risks and Responses*, Health and Global Environmental Change Series No. 2, 2004.

▌ World Health Organization. *Millennium Ecosystem Assessment. Ecosystems and Human Well-Being: A Health Synthesis*, 2005.

▌ World Resources Institute. *World Resources 2005: The Wealth of the Poor: Managing Ecosystems to Fight Poverty*, World Resources Institute (WRI) in collaboration with United Nations Development Programme, United Nations Environment Programme and The World Bank, 2005.

▌ Yande, Dai, Yuezhong, Zhu & Sint, Jonathan E. 'China's Energy Demand Scenarios to 2020', *The Sinosphere Journal*, vol. 7, Issue 1, May 2004.

▌ Zhi Dong, Li. *Energy and Environmental Problems Behind China's High Economic Growth– A Comprehensive Study of Medium- and Long-term Problems, Measures and International Cooperation*, The Institute of Energy Economics, Japan, March 2003.

5. SCIENCE & TECHNOLOGY

▌ Altmann, Jürgen. 'Military Uses of Nanotechnology: Perspectives and Concerns', *Security Dialogue*, vol. 35, no. 1, pp. 61-79, March 2004.

▌ Anton, Philip S., et al. *The Global Technology Revolution*, RAND MR-1307 (Santa Monica, California, 2001).

▌ Coates, Joseph F., et al. 2025: *Scenarios of US and Global Society Reshaped by Science and Technology* (Oakhill Press, Greensboro, 1998).

▌ 'Creating an Innovative Europe', Report of the Independent Expert Group on R&D and Innovation appointed following the Hampton Court Summit, January 2006.

▌ Dubash, Manek. 'Moore's Law is dead says Gordon Moore', *Techworld*, 13 April 2005.

▌ Fisher, Brian S., et al. *Technological development and economic growth*, Abare Research Report, January 2006.

▌ Foster, Ian. 'The Grid: Computing without Bounds', *Scientific American*, vol. 288, no. 4, April 2003.

▌ The Freedonia Group. 'Nanotechnology in Health Care', May 2005, accessed through *Nanotechnology Now:* http://www.nanotech-now.com/news.cgi?story_id=09445 (30 January 2006).

▌ Holz, Carsten A. 'China's Economic Growth 1978-2025: What We Know Today about China's Growth Tomorrow', Centre on China's Transnational Relations, Working Paper no. 8, The Hong Kong University of Science and Technology, July 2005.

▌ Hundley, Richard O., et al. *The Global Course of the Information Revolution: Recurring Themes and Regional Variations*, RAND MR-1680-NIC (Santa Monica, California, 2003).

- Institute of Higher Education, Shanghai Jiao Tong University. *Academic Rankings of World Universities 2005*, 2005.

- Intel Corporation. 'Platform 2015: Intel Processor and Platform Evolution for the Next Decade', Intel White Paper, 2005.

- Kanellos, Michael. 'Intel sketches out nanotechnology road map', *news.com*, 25 October 2005.

- Lerner, Preston. 'Robots go to war', *Popular Science*, vol. 268, no. 1, January 2006.

- *Nanotechnologies, societal implications – maximizing benefits for humanity*, Report of the National Nanotechnology Initiative Workshop, 2-3 December 2003.

- National Nanotechnology Initiative. Available at: http://www.nano.gov/html/res/faqs.html

- New Zealand Ministry of Research, Science and Technology. 'Biotechnologies to 2025', Report prepared for the New Zealand Agencies by the Ministry of Research, *Science and Technology*, January 2005.

- Pollack, Andrew. 'Genetically-engineered Crops: The Next Generation', *International Herald Tribune*, 15 February 2006.

- Rader, Michael, et al. 'Key Factors Driving the Future Information Society in the European Research Area', Technical Report Series, European Commission, Institute for Prospective Technological Studies, Joint Research Centre, September 2004.

- Teresko, John. 'Get Ready for the Age of Nanotechnology', *Forbes.com*, October 2003.

- 'The National Nanotechnology Initiative: Research and Development Leading to a Revolution in Technology and Industry', Supplement to the President's FY 2006 Budget, March 2005.

- UNESCO, *UNESCO Science Report 2005*, Paris 2005.

- World Intellectual Property Organization. 'Exceptional Growth from North East Asia in Record Year for International Patent Filings', Press Release 436, Geneva, 3 February 2006.

- Zhou, Ping & Leydesdorff, Loet. *The Emergence of China as a Leading Nation in Science*, Forthcoming policy paper.

PART II: The Regions

Also Final Section: **The EU in Context**

- *A secure Europe in a better world*, European Security Strategy, 12 December 2003 (accessible via http://www.iss.europa.eu/solana/solanae.html).

- Abélès, Marc. *Politique de la survie* (Flammarion, Paris, 2006).

- Abou Zahab, Mariam. 'Pakistan : entre l'implosion et l'éclatement?', *Politique étrangère*, no. 2, IFRI, Paris, 2006.

- Achcar, Gilbert. *Maxime Rodinson on Islamic 'Fundamentalism'*, Middle East Report, vol. 34, no. 233, Winter 2004.

- Ahya, Chetan & Sheth, Mihir. *India. Infrastructure: Changing Gears*, Morgan Stanley Report, 2005.
- Alam, Asad, et al. *Growth, Poverty and Inequality in Eastern Europe and the Former Soviet Union*, World Bank Report, October 2005.
- Amato, Giuliano (with Lucia Pozzi). *Un altro mondo è possibile* (Arnoldo Mondadori Editore, Milan, 2006).
- Andrews-Speed, Philip. 'China's energy policy and its contribution to international stability', in Zaborowski, Marcin (ed.), *Facing China's Rise: Guidelines for an EU Strategy*, Chaillot Paper (EUISS, Paris, forthcoming).
- Ayres, Robert. *Crime and Violence as Development Issues in Latin America and the Carribean* (The World Bank, Washington D.C., 1998).
- Barisitz, Stephan. *Distorted Incentives Fading? The Evolution of the Russian Banking Sector since Perestroika*, Central Bank, Republic of Austria, 2004.
- Barkey, Henri J. & Fuller, Graham. *Turkey's Kurdish Question* (Rowman and Littlefield, New York, 1998).
- Bayat, Kaveh. *The Ethnic Question in Iran*, Middle East Report, no. 237, Winter 2005.
- Beck, Ulrich. *Risk Society: Towards a New Modernity* (Sage, London, 1992).
- Bergen, Peter. *Holy War Inc.: Inside the Secret World of Osama bin Laden* (Free Press, New York, 2002).
- Bergsten, C. Fred et al. *China: The Balance Sheet. What the World Needs to Know Now about the Emerging Superpower*, Center for Strategic and International Studies and Institute for International Economics (Public Affairs, New York, 2006).
- Bertrand, Gilles, Michalski, Anna & Pench, Lucio. *European Futures: Five Possible Scenarios for 2010* (Edward Elgar Publishing Ltd, Cheltenham, 2000).
- Bijian, Zhen. '"Peacefully Rising" to Great-Power Status', *Foreign Affairs*, vol. 84, no. 5, September/October 2005.
- Biscop, Sven. *The European Security Strategy: A Global Agenda for Positive Power* (Ashgate, Aldershot, 2005).
- Boquérat, Gilles. 'Une puissance en quête de reconnaissance', *Questions internationales*, no. 15, La Documentation française, Paris, September/October 2005.
- Bozarslan, Hamit. *La question kurde. États et minorités au Moyen-Orient* (Presses de Sciences Po, Paris, 1997).
- Braud, Pierre-Antoine. 'La Chine en Afrique: Anatomie d'une nouvelle stratégie chinoise', EUISS, Paris, October 2005.
- Brunner, Rainer & Ende, Werner (eds.) *The Twelver Shia in Modern Times. Religious Culture and Political History* (Brill, Leiden, 2001).
- Bueno de Mesquita, Bruce & Downs, George. 'Development and Democracy', *Foreign Affairs*, vol. 84, no. 5, September/October 2005.
- 'Can India Fly?', Special Report, *The Economist*, 1 June 2006.

Bibliography

- Carter, Ashton. 'America's New Strategic Partner?' *Foreign Affairs*, vol. 85, no. 4, July/August 2006.

- Chaliand, Gérard (ed.) *A People without a Country. The Kurds and Kurdistan* (Zed Books, London, 1980, several reeditions).

- Chauvin, Sophie & Lemoine, Françoise. *L'économie indienne: changements structurels et perspectives à long terme*, CEPII Working Paper no. 2005-04, April 2005.

- Chellaney, Brahma. 'India's Future Security Challenge: Energy Security', in *India as a New Global Leader* (The Foreign Policy Centre, 2005).

- Chinese Ministry of Environment. *Report on the State of the Environment In China*, 2004.

- Cleuntinx, Christian. *The EU-Russia Energy Dialogue* (DG for Energy and Transport, European Commission, Vienna, December 2003).

- Cline, William R. *The Case for a New Plaza Agreement, Policy Briefs in International Economics*, Institute for International Economics, December 2005.

- Cooper, Robert. *The Breaking of Nations: Order and Chaos on the 21st Century* (Atlantic Books, London, 2003).

- Council on Foreign Relations (CFR). *Russia's Wrong Direction: What the United States Can and Should Do*, March 2006.

- Daalder, Ivo H, Gnesotto, Nicole & Gordon, Philip H. (eds.) *Crescent of Crisis: US-European Strategies for the Greater Middle East* (Brookings Institution Press & EU Institute for Security Studies, 2006).

- Das, Gurcharan. 'The India Model', *Foreign Affairs*, vol. 85, no. 4, July/August 2006.

- de Mesquitas Neto, Paulo. Crime, *Violence and Democracy in Latin America*, Paper submitted for the Integration in the Americas Conference, 2 April 2002.

- Dean, Andrew. *Challenges for the Russian Economy*, OECD Economics Department, Moscow, 7 July 2004.

- Deutsche Bank Research. *Dynamic Sectors give global growth centres the edge*, Current Issues, October 2005.

- Dillon, Michael. *Global Security in the 21st Century. Circulation, Complexity and Contingency*, Chatam House Briefing Paper 05/02, October 2005.

- Economist Intelligence Unit. *Foresight 2020: Economic, Industry and Corporate Trends*, 2006.

- 'Energy Policy Act 2005, Section 1837: National Security Review of International Energy Requirements', US Department of Energy, February 2006.

- Energy Research Institute. *China National Energy Strategy and Policy to 2020, Subtitle 2: Scenario Analysis on Energy Demand*, National Development and Reform Commission, China Sustainable Energy Program, May 2004.

- Esposito, John L. *Unholy Wars: Terror in the Name of Islam* (Oxford University Press, New York, 2002).

- European Commission. *A Stronger Partnership between the European Union and Latin America*, 8 December 2005, COM (2005) 636 final.

- European Round Table of Industrialists. *Seizing the Opportunity: Taking the EU-Russia Relationship to the Next Level*, ERT, May 2006.
- Faath, Sigrid (ed.) *Politische und gesellschaftliche Debatten in Nordafrika, Nah- und Mittelost. Inhalte, Träger, Perspektiven* (DOI Mitteilungen 72, Hamburg 2004).
- Farrell, Mary, et al. (eds.) *Global Politics of Regionalism* (Pluto Press, Cambridge, 2004).
- Fay, Marianne & Morrison, Mary. *Infrastructure in Latin America & the Caribbean: Recent Developments and Key Challenges*, The World Bank, Washington D.C., 2005.
- Featherstone, Mike, et al. *Global Modernities* (Sage, London, 1995).
- Featherstone, Mike. *Global Culture: Nationalism, Globalisation and Modernity* (Sage, London, 1990).
- Ferguson, James. *The Anti-Politics Machine: 'development', depoliticization and bureaucratic power* (University of Minneapolis Press, Minneapolis, 1994).
- Ferguson, Niall. 'A World Without Power', *Foreign Policy*, July/August 2004.
- Food and Agriculture Organization of the United Nations. *Poverty Alleviation and Food Security in Asia: Lessons and Challenges*, FAO Regional Office for Asia and the Pacific, December 1998.
- Foreign Investment Advisory Council. *Russia: Investment Destination*, March 2005.
- Forster, Christopher J. *China's Secret Weapon? Science Policy and Global Power*, The Foreign Policy Centre, 2006.
- Freedman, Lawrence. *The Transformation of Strategic Affairs*, Adelphi Paper 379, International Institute for Strategic Studies (IISS), 2006.
- Freedom House. *Freedom in the World 2006*, 2006.
- Friedman, Thomas. *The World is Flat: A Brief History of the Twenty-first Century* (Farrar, Straus and Giroux, New York, 2005).
- Fukuyama, Francis. *After the Neocons: America at the Crossroads* (Yale University Press, New Haven, 2006).
- Fuller, Graham E. & Rahim Francke, Rend. *The Arab Shi'a. The Forgotten Muslims* (St. Martin's Press, New York, 1999).
- Galeotti, Mark (ed.) *Russian and Post-Soviet Organized Crime* (Ashgate, Aldershot, 2002).
- Gayer, Laurent. 'Conflits et coopérations régionales en Asie du Sud', *Questions internationales*, no. 15, La Documentation française, Paris, September/October 2005.
- Gerges, Fawaz A. *The Far Enemy: Why Jihad Went Global* (Cambridge University Press, New York, 2005).
- *Globalization of the Media Industry and Possible Threats to Cultural Diversity*, Final Study, European Parliament – Directorate General for Research, July 2001.
- Gnesotto, Nicole (ed.). *EU Security and Defence Policy — The First Five Years (1999-2004)* (EUISS, Paris, 2004).
- Goldsmith, Benjamin. *Here there be Dragons: The Shanghai Cooperation Organisation*, Centre for Defence Information, 2005.

Bibliography

- Gordon, Philip H. 'America's Role in the World: Searching for Balance', in Zaborowski, Marcin (ed.) *Friends Again? EU-US relations after the crisis* (EUISS, Paris, 2006).

- Gowan, Richard. *The EU, regional organisations and security: strategic partners or convenient alibis?*, Egmont Paper 3, Royal Institute for International Relations (IRRI-KIIB), 2004.

- Greenberg, Karen J. *Al Qaeda Now. Understanding Today's Terrorists* (Cambridge University Press, New York, 2005).

- Griswold, Wendy. *Cultures and Societies in a Changing World* (Sage, London, 2003).

- Halliday, Fred. *The Middle East in International Relations: Power, Politics and Ideology* (Cambridge University Press, New York, 2005).

- Halm, Heinz. *Die Schia* (Wissenschaftliche Buchgesellschaft, Darmstadt, 1988).

- Held, David. *Global Covenant: The Social Democratic Alternative to the Washington Consensus* (Polity Press, Cambridge, 2004).

- Hourcade, Bernard. 'Iran's internal security challenges,' in Posch, Walter (ed.) *Iranian Challenges*, Chaillot Paper no. 89 (EUISS, Paris, May 2006).

- Human Security Centre. *Human Security Report 2005: War and Peace in the 21st Century*, HSC, The University of British Columbia, 2005.

- Huntington, Samuel P. 'The Hispanic Challenge', *Foreign Policy*, March/April 2004.

- Ikenberry, G. J. 'A Weaker World', *Prospect*, November 2005.

- International Crisis Group. *Youth in Central Asia: Losing the New Generation*, Asia Report no. 66, October 2003.

- International Crisis Group. *Is Radical Islam inevitable in Central Asia? Priorities for Engagement*, Asia Report no. 72, December 2003.

- Interview with Sunil Khilnani, 'La société indienne: tensions et transformations', *Questions internationales*, no. 15, La Documentation française, Paris, September/October 2005.

- Jaffrelot, Christophe. 'Inde: un tropisme américain aux dépens de l'Europe ?', *Le Monde diplomatique*, September 2005.

- Jaffrelot, Christophe. 'L'Inde, la puissance pour quoi faire?', *Politique internationale*, October 2006.

- Kaldor, Mary. *Global Civil Society: An Answer to War* (Polity Press, Cambridge, 2003).

- Kepel, Gilles & Milleli, Jean-Pierre. *Al-Qaida dans le texte* (Presses Universitaires de France, Paris, 2005).

- Kepel, Gilles. *Fitna: Guerre au Cœur de l'Islam* (Gallimard, Paris, 2004).

- Kepel, Gilles. *Jihad: Expansion et déclin de l'islamisme* (Gallimard, Paris, 2000).

- Kirton, John. *From Collective Security to Concert: The UN, G8 and Global Security Governance*, paper presented at conference on 'Security Overspill; Between Economic Integration and Social Exclusion', Montreal 2005.

- Kirton, John. *From G7 to G20: Capacity, Leadership and Normative Diffusion in Global Financial Governance*, paper presented at International Studies Association Annual Convention, Hawaii, 1-5 March 2005.

- Kirton, John. *The Future of the G8*, paper submitted at the Gleneagles summit, Scotland, July 2005.

- Kohut, Andrew & Stokes, Bruce. *America Against the World: How We are Different* (Times Books, New York, 2006).

- Krasner, Stephen D. (ed.). *International Regimes* (Cornell University Press, New York, 1983).

- Kristol, William & Kagan, Robert. 'Towards a Neo-Reaganite Foreign Policy', *Foreign Policy*, July/August 1996.

- Kupchan, Charles. *The End of the American Era: US Foreign Policy and the Geopolitics of the Twenty-First Century* (Random House, New York, 2002).

- 'L'energia al potere', *Aspenia*, no. 32, 2006.

- Ladier-Fouladi, Marie. 'Population et politique en Iran. De la monarchie à la République islamique', *Les Cahiers de l'INED*, no. 150, Institut national d'études démographiques, Paris, 2003.

- Laïdi, Zaki. *La norme sans la force. L'énigme de la puissance européenne* (Presses de Sciences Po, Paris, 2005).

- Lamy, Pascal. *The Emergence of Collective Preferences in International Trade: Implications for Regulating Globalisation* (PROSES, Sciences Po, 2004).

- Langohr, Vickie. *Experiments in Multi-Ethnic and Multi-Religious Democracy*, Middle East Report, vol. 35, no. 237, Winter 2005.

- Leijonhielm, Jan & Larsson, Robert L. *Russia's Strategic Commodities: Energy and Metals as Security Levers*, Swedish Defence Research Agency, November 2004.

- Leonard, Mark. *Why Europe Will Run the 21st Century* (Fourth Estate, London, 2005).

- Lieven, Anatol. *America Right or Wrong: An Anatomy of American Nationalism* (Oxford University Press, Oxford, 2004).

- Lindert, Kathy, Skoufias, Emmanuel & Shapiro, Joseph. *Redistributing Income to the Poor and the Rich: Public Transfers in Latin America and the Caribbean*, The World Bank, Washington D.C., 30 March 2006.

- Luciani, Giacomo. 'Oil and Political Economy in the International Relations of the Middle East,' in Fawcett, Louise (ed.), *International Relations of the Middle East* (Oxford University Press, 2005).

- Lynch, Dov. *Engaging Eurasia's Separatist States. Unresolved Conflicts and De Facto States* (United States Institute of Peace, Washington D.C., 2004).

- Lynch, Dov (ed.). *What Russia Sees*. Chaillot Paper no. 74 (EUISS, Paris, January 2005).

- Marat, Erica. 'Impact of Drug Trade and Organized Crime on State Functioning in Kyrgyzstan and Tajikistan,' *China and Eurasia Forum Quarterly*, vol. 4, no. 2, 2006.

- Marty, Martin & Applebee, Scott. *Fundamentalisms and Society: Reclaiming the Sciences, the Family, and Education* (University Of Chicago Press, Chicago, 1997).

I McDowall, David. *A Modern History of the Kurds* (I.B. Tauris, London, 1997).

I Mearsheimer, John J. *The Tragedy of Great Power Politics* (W.W. Norton and Company, New York, 2001).

I Mistral, Jacques & Salzman, Bernard. 'La préférence américaine pour l'inégalité', *En Temps Réel*, Cahier 25, 2006.

I Mitchell, Richard P. *The Society of the Muslim Brotherhood* (Oxford University Press, 1969, republished 1993).

I Mohan, C. Raja. 'India and the Balance of Power', *Foreign Affairs*, vol. 85, no. 4, July/August 2006.

I Mousalli, Ahmad S. *Islamic Fundamentalism: Myths and Realities* (Ithaca, Reading, 1998).

I Myers, Norman. 'Environmental refugees: an emergent security issue', 13th Economic Forum, Session III – Environment and Migration, Prague, 23-27 May 2005.

I Nankivell, Nathan. *The National Security Implications of China's Emerging Water Crisis*, China Brief, vol. 5, no. 17, The Jamestown Foundation, 2 August 2005.

I Nasr, Mamdouh. 'Assessing Desertification in the Middle East and North Africa: Policy Implications', in Brauch, Hans Günter et al. (eds.), *Security and the Environment in the Mediterranean: Conceptualising Security and Environmental Conflicts* (Springer Verlag, Berlin-Heidelberg, 2003).

I National Drug Control Strategy of the United States. FY 2006 Budget Summary, February 2005.

I National Drug Intelligence Center. *National Drug Threat Assessment 2006* (NDIC, Washington D.C., January 2006).

I National Intelligence Council. *Mapping the Global Future: Report of the National Intelligence Council's 2020 Project*, December 2004.

I Nye Jr., Joseph S. *Soft Power: The Means to Success in World Politics* (Public Affairs, New York, 2004).

I O'Leary, Brendan et al. (eds.) *The Future of Kurdistan in Iraq* (University of Pennsylvania Press, Philadelphia, 2005).

I Olson, Robert (ed.) *The Kurdish Nationalist Movement in the 1990s: Its Impact on Turkey and the Middle East* (University Press of Kentucky, Lexington, 1996).

I Olson, Robert. *The Kurdish Question and Turkish-Iranian Relations. From World War I to 1998* (Mazda Publishers, Costa Mesa, 1998).

I Oppenheimer, Andrés. *Cuentos chinos. El engaño de Washington, la mentira populista y la esperanza de América Latina* (Sudamericana, Buenos Aires, 2005).

I Organisation for Economic Co-operation and Development. *OECD Science, Technology and Industry Scoreboard 2005 – Towards a knowledge-based economy*, October 2005.

I Ortega, Martin (ed.). *Global Views on the European Union*, Chaillot Paper no. 72 (EUISS, Paris, November 2004).

I Ortega, Martin (ed.). *The EU and the UN: Partners in Effective Multilateralism*, Chaillot Paper no. 78 (EUISS, Paris, June 2005).

- Ortega Carcelén, Martin. *Cosmocracia, Politica global para el siglo XXI* (Editorial Sintesis, Madrid, 2006).
- Paillard, Christophe-Alexandre. 'L'Amérique latine, nouvel acteur majeur du grand jeu énergétique mondial', *Défense nationale*, no. 4, April 2006.
- Pei, Minxin. 'The Dark Side of China's Rise', *Foreign Policy*, March/April 2006.
- Pew Research Centre for the People and the Press. *2002 Global Attitudes Survey*, Washington D.C., April 2002.
- Pew Research Center for the People and the Press. *Survey of Foreign Policy*, Washington D.C., July 2004.
- Pew Hispanic Centre. 'Unauthorized Migrants: Numbers and Characteristics', Washington D.C., 14 June 2005.
- Pew Research Center for the People and the Press. *Survey of Religion*, Washington D.C., July 2005.
- Pew Research Center for the People and the Press. *Foreign Policy Attitudes*, Washington D.C., July 2005.
- Pew Research Centre for the People and the Press. *15-Nation Pew Global Attitudes Survey*, Washington D.C., June 2006.
- Phillips, Kevin. *American Theocracy: The Peril and Politics of Radical Religion, Oil, and Borrowed Money in the 21st Century* (Viking, New York, 2006).
- Posch, Walter (ed.). *Looking into Iraq*, Chaillot Paper no. 79 (EUISS, Paris, July 2005).
- Poulton, Hugh. *Top Hat, Grey Wolf and Crescent: Turkish Nationalism and the Turkish Republic* (New York University Press, New York, 1997).
- Ramo, Joshua Cooper. *The Beijing Consensus*, The Foreign Policy Centre, London, 2004.
- Richards, Alan. 'Democracy in the Arab Region: Getting There from Here,' in *Middle East Policy*, vol. XII, no. 2, Summer 2005.
- Rischard, Jean François. *High Noon: 20 global problems and 20 years to solve them* (Perseus Book Group, New York, 2002).
- Round table discussion: 'Democracy: Rising Tide or Mirage?' in *Middle East Policy*, vol. XII, no. 2, Summer 2005.
- Roy, Olivier. *Globalized Islam: The Search for a New Ummah* (Columbia University Press, New York, 2004).
- Ruggie, John Gerard. *Multilateralism Matters: The Theory and Praxis of an Institutional Form* (Columbia University Press, New York, 1993).
- Sachs, Jeffrey D. (dir.). *Investing in Development, A Practical Plan to Achieve the Millennium Development Goals* (UN Millennium Project, 2005).
- Sageman, Marc. *Understanding Terror Networks* (University of Pennsylvania Press, Philadelphia, 2004).
- Salamé, Ghassan (ed.). *Democracy without Democrats? The Renewal of Politics in the Muslim World* (I.B. Tauris, London-New York, 1994).

Bibliography

- Sander, Cerstin & Munzele Maimbo, Samuel. *Migrant Labor Remittances in Africa:Reducing Obstacles to Developmental Contributions*, Africa Region Working Paper Series no. 64, The World Bank, Washington D.C., 2003.

- Schwedler, Jillian & Chomiak, Laryssa. *And the Winner is... Authoritarian Elections in the Arab World*, Middle East Report, vol. 36, no. 238, Spring 2006.

- Sen, Amartya. 'Contrary India', in *The Economist: The World in 2006* (The Economist Newspaper Limited, London, 2005).

- Seufert, Günter. *Politischer Islam in der Türkei. Islamismus als Symbolische Repräsentation einer sich modernisierenden muslimischen Gesellschaft* (Beiruter Texte und Studien 67/Türkische Welten 5, Istanbul/Stuttgart, 1997).

- Silverstein, Paul. *State and Fragmentation in North Africa*, Middle East Report, no. 237, Winter 2005.

- Slaughter, Anne Marie. *A New World Order* (Princeton University Press, NJ, 2004).

- Small, Andrew. *Preventing the Next Cold War: A View from Beijing*, The Foreign Policy Centre, London, 2005.

- 'Special Report: Inequality in America', *The Economist*, 17 June 2006.

- Stansfield, Gareth. *Iraqi Kurdistan: Political Development and Emergent Democracy* (Routledge, London, 2003).

- Strohmeier, Martin & Yalçin-Heckmann, Lale. *Die Kurden. Geschichte, Politik, Kultur* (C.H.Beck, Munich, 2000).

- Sullivan, Mark P. *Latin America: Terrorism Issues*, CRS Report for Congress RS21049, Congressional Research Service, Washington D.C., 18 January 2006.

- Symposium notes: 'A Shia Crescent? What Fallout for the US?' in *Middle East Policy*, vol. XII, no. 4, Winter 2005.

- Teló, Mario (ed.). *European Union and New Regionalism* (Ashgate, Aldershot, 2001)

- Teló, Mario. *Europe: A Civilian Power? European Union, Global Governance, World Order* (Palgrave Macmillan, Basingstoke, 2006).

- Terrill, Ross. *The New Chinese Empire* (Basic Books, New York, 2003).

- The World Bank. *Russian Economic Report*, Moscow Office, Economics Unit, April 2006.

- The World Trade Organization. *The Future of the WTO – Addressing institutional challenges in the new millennium*, Report by the Consultative Board to the Director-General Supachai Panitchpakdi, Geneva, 2004.

- The World Trade Organization. *International Trade Statistics 2005*, Geneva, 2005.

- Thomas, Ward. 'La légitimité dans les relations internationales: dix propositions', in: Andréani, Gilles and Hassner, Pierre (eds.), *Justifier la guerre? De l'humanitaire au contre-terrorisme* (Presses de Sciences Po, Paris, 2005).

- Timperlake, Edward & Triplett III, William C. *Red Dragon Rising: Communist China's Military Threat to America* (Regnery, Washington D.C., 2002).

- Transparency International. *Global Corruption Report 2006*.

- Treverton, Gregory F. et al., *Exploring Religious Conflict* (RAND, Santa Monica, 2005).

- UNAIDS. *HIV/AIDS. Country Profile: India*, Population Division, Department of Economic and Social Affairs, United Nations, March 2003.

- United Nations. *Millennium Development Goals*. Available at: http://www.un.org/millennium goals/.

- United Nations. *World Economic and Social Survey: International Migration*, Population Division, Department of Economic and Social Affairs, 2004.

- United Nations Development Programme. *Cultural Liberty in Today's Diverse World*, UN Human Development Report, 2004.

- United Nations Development Programme. *Human Development Report 2005*, 2005.

- United Nations Office on Drugs and Crime. *2005 World Drug Report, Part 1*, 2005.

- US National Intelligence Council. *Latin America 2020: Discussing Long-Term Scenarios*, Summary of conclusions of the workshop on Latin American trends, Santiago de Chile, 7-8 June 2004.

- Uvin, Peter. *Aiding Violence: The Development Entreprise in Rwanda* (Kumarian Press, West Hartford, 1998).

- van Bruinessen, Martin. *Agha, Sheikh and State. The Social and Political Structures of Kurdistan* (Zed Books, London 1992).

- Védrine, Hubert. *Face à l'hyperpuissance: textes et discours 1995-2003* (Fayard, Paris, 2003).

- Williamson, John. *What Follows the USA as the World's Growth Engine?*, India Policy Forum, India Policy Forum Public Lecture, 25 July 2005.

- Yavuz, M. Hakan & Özcan, Nihat Ali. 'The Kurdish Question in Turkey,' in *Middle East Policy*, vol. XIII, no. 1, Spring 2006.

- Zaborowski, Marcin. 'US China Policy: Implications for the EU', EUISS, Paris, October 2005.

- Zakaria, Fareed. 'The Rise of Illiberal Democracy,' *Foreign Affairs*, vol. 76, no. 6, November/December 1997.

- Zakaria, Fareed. *The Future of Freedom: Illiberal Democracy at Home and Abroad* (W.W. Norton & Company, New York, 2003).

- ZhiDong, Li. *Energy and Environmental Problems behind China's High Economic Growth – A Comprehensive Study of Medium- and Long-term Problems, Measures and International Cooperation*, Institute of Energy Economics (IEEJ), Japan, March 2003.

- Zins, Max-Jean. 'La plus grande démocratie du monde?', *Questions internationales*, no. 15, La Documentation française, Paris, September/October 2005.

Abbreviations

AIDS	Acquired Immune Deficiency Syndrome
AKP	Justice and Development Party (Adalet ve Kalkýnma Partisi)
ASEAN	Association of South-East Asian Nations
AU	African Union
b/d	barrels a day
bcm	billion cubic metres
BIMST-EC	Bangladesh, India, Myanmar, Sri Lanka, Thailand Economic Cooperation organisation
BRIC	Brazil, Russia, India and China
btoe	billion tonnes of oil equivalent
cf	cubic feet
CIS	Confederation of Independent States
cm	cubic metres
CTBT	Comprehensive Test Ban Treaty
EBRD	European Bank for Reconstruction and Development
ECOWAS	Economic Community of Western African States
EIA	Energy Information Administration
ENSO	El Niño Southern Oscillation
ETS	Emissions Trading Scheme
FDI	Foreign Direct Investment
GCC	Gulf Cooperation Council
GDP	Gross Domestic Product
GERD	Gross Expenditure on Research and Development
GHG	Greenhouse gases
GW	gigawatts
HIV	Human Immunodeficiency Virus
IAEA	International Atomic Energy Agency
ICT	Information and Communication Technology
IDP	Internally Displaced Person
IEA	International Energy Agency
IMF	International Monetary Fund
inh/km²	Inhabitants per square kilometre
IT	Information Technology
kWh	kilowatt-hour(s)
LNG	Liquefied Natural Gas
mb/d	million barrels a day
MDG	Millennium Development Goals
MENA	The Middle East and North Africa

Mt	million tonnes
NAFTA	North American Free Trade Agreement
NATO	North Atlantic Treaty Organisation
NEPAD	New Partnership for Africa's Development
NPT	Treaty on the Non-Proliferation of Nuclear Weapons
NT	Nanotechnology
OECD	Organisation for Economic Co-operation and Development
OPEC	Organisation of the Petroleum Exporting Countries
ppmv	parts per million by volume
PPP	Purchasing Power Parity
R&D	Research and Development
RFA	Regional Financial Arrangement
S&T	Science and Technology
SAARC	South Asian Association for Regional Cooperation
SAFTA	South Asian Free Trade Agreement
SARS	Severe Acute Respiratory Syndrome
TB	Tuberculosis
UAE	United Arab Emirates
UIA	United Iraqi Alliance
UN	United Nations
UNDP	United Nations Development Programme
UNEP	United Nations Environment Programme
UNHCR	United Nations High Commissioner for Refugees
UNSC	United Nations Security Council
WHO	World Health Organization
WTO	World Trade Organization